Strategies For Short-Term Testing For Mutagens/Carcinogens

Editor

Byron E. Butterworth

Chief of Genetic Toxicology
Chemical Industry Institute of Toxicology
Research Triangle Park, North Carolina

Editor-in-Chief
CRC Toxicology Series

Leon Golberg

President
Chemical Industry Institute of Toxicology
Research Triangle Park, North Carolina

CRC Press, Inc.
2255 Palm Beach Lakes Blvd. · West Palm Beach, Florida 33409

QH
465
C5
S77

Library of Congress Cataloging in Publication Data

Main entry under title:

Strategies for short-term testing for mutagens/car-
  cinogens.
  (CRC toxicology series) (Uniscience series on
toxicology)
  Proceedings of the Chemical Industry Institute of
Toxicology workshop, Research Triangle Park, N. C.,
Aug. 11-12, 1977.
  Bibliography: p.
  Includes index.
  1. Chemical mutagenesis—Congresses. 2. Car-
cinogens—Testing—Congresses. I. Butterworth,
Byron E. II. Chemical Industry Institute of Toxicology. III. Series: Chemical Rubber Company,
Cleveland. CRC toxicology series.
QH465.C5S77   575.2'92   78-12587
ISBN 0-8493-5661-X

This book represents information obtained from authentic and highly regarded sources. Reprinted material is quoted with permission, and sources are indicated. A wide variety of references are listed. Every reasonable effort has been made to give reliable data and information, but the author and the publisher cannot assume responsibility for the validity of all materials or for the consequences of their use.

All rights reserved. This book, or any parts thereof, may not be reproduced in any form without written consent from the publisher.

© 1979 by CRC Press, Inc.

International Standard Book Number 0-8493-5661-X

Library of Congress Card Number 78-12587
Printed in the United States

PREFACE

Many new short-term tests have been proposed to predict the mutagenic/carcinogenic activity of chemicals, and it would appear that these tests have the potential of revolutionizing this area of toxicology. However, there currently exists a wide range of opinion as to the usefulness of the short-term tests. On the one hand, some investigators feel that these assays are so well developed and validated that they should be adopted by regulatory agencies as a valuable part of the health hazard assessment of chemicals. They believe that the testing procedures and interpretation of results can be rigorously defined in regulations so that governmental administrators and lawyers can fairly and usefully apply such laws. On the other hand, others see the tests as having some value, but being capable of producing misleading information. They consider it doubtful, at best, that the ability of a chemical to induce mutagenic effects in bacteria could in any way realistically predict the role such a chemical might play in the complex process of carcinogenesis in man. They point out that for many proposed short-term assays experimental details have not been well defined, important factors in the tests such as the metabolic activation system are poorly understood and lack standardization, and most tests have not been evaluated with reasonable numbers of compounds of known mutagenicity/carcinogenicity.

The goal of the Chemical Industry Institute of Toxicology Workshop on "Strategies for Short-Term Testing for Mutagens/Carcinogens" was to provide an assessment of the status and usefulness of the major short-term tests and to produce a set of practical guidelines for dealing with the results of these assays. The conference was attended by over 50 experts from academia, government, and industry active in this field. The account of the workshop begins with a section comprising authoritative and critical reviews of a number of the more promising assays. This section is followed by a discussion of a proposed set of practical guidelines for using and interpreting the results of the tests. Finally, and this was a unique feature of the workshop, there were discussions of several hypothetical test cases depicting the kinds of situations that arise when attempting to formulate a proper course of action in response to the results of various short-term tests.

Obviously, with such a complex area there was not unanimity on every point. In fact, the wide variety of opinions expressed illustrates the state of flux of the field. There was, however, general agreement on several major points. It was clear that a unique set of circumstances will surround each chemical that is subjected to these tests and that each chemical should be evaluated on a case-by-case basis. It was felt that, where possible, testing should be done with a battery of different type of tests. At this stage of development, it would be unwise to base decisions on the results of any single test. Of course, each testing program must be independently formulated to meet specific needs. Over 40 different tests were reviewed. Before reliance is placed on the results of a test, the validation studies should be examined to see if the particular test in question responds properly to the particular structural class of chemical in question. It was felt that statements such as: "the results in this test correlate 90% with animal carcinogenicity studies" were misleading because the measure of correlation can be manipulated by judicious selection of the chemicals tested. A further suggestion in formulating a battery of tests would be to incorporate at least one system that involves the use of mammals or mammalian cells. Also, many investigators felt that cell transformation systems showed the most promise as accurate predictors of the carcinogenic potential of chemicals.

The suggestions that emerged from the workshop for formulating a testing program may be considered by many to be somewhat idealistic. It was pointed out that much more experimentation needs to be done to define most systems. The metabolic activation portion of the tests was of particular concern. Extrapolation from metabolism in a rat liver homogenate to the complex metabolism in man is risky and further research in this area is indicated. Another important consideration is that the resources and expertise necessary for accurately performing many of the tests are severely limited, especially with some of the more valuable advanced tests. For example, initial results with the hamster embryo cell transformation assay indicate that it will be an accurate predictor of carcinogenic potential for many classes of chemicals.* The test requires a specially equipped laboratory for maintaining cells in culture and handling hazardous chemicals. Defined procedures must be followed rigorously, and even commonly accepted deviations in technique can render the test invalid. It was estimated that it would take 6 months to a year for a laboratory already familiar with cell culture to successfully run the hamster embryo cell transformation assay on a routine basis. Currently, there exists only a handful of investigators who are qualified to perform this test.

The workshop concluded that results from a battery of validated tests could be of significant value in assessing the carcinogenic potential of a chemical, provided that the overall assessment was made in the context of a rational examination of all the facts concerning the compound. Positive results in several valid short-term tests indicate that, without waiting for the results of long-term animal exposure studies, operations involving the chemical should be immediately examined and human exposure reduced to as far as is practicable. Unfortunately, those responsible are left with little guidance as to what acceptable exposure levels might be, because the short-term tests are not developed sufficiently to predict potency. The results of lifetime animal exposure studies would be extremely valuable in confirming carcinogenic potential of a chemical and in establishing acceptable safe limits of exposure, and such studies should be considered. Nevertheless, positive results from an adequate short-term testing battery cannot be ignored and an effective program to protect people who are likely to be exposed should be instituted without delay.

B. E. Butterworth

---

* Pienta, R. et al., *Int. J. Cancer*, 19, 642, 1977.

INTRODUCTION

Many short-term tests have been proposed to predict the carcinogenic activity of chemicals. In spite of the limitations of our current state of knowledge about the existing tests, many are in use and are in the course of being adopted by regulatory agencies in various parts of the world. There is currently little practical guidance as to which tests are relevant and valid and "what to do next" as the results of the tests appear. The goal of the Chemical Industry Institute of Toxicology (CIIT) Workshop on *Strategies for Short-Term Testing for Mutagens/Carcinogens* was to provide an assessment of the status and usefulness of the major short-term tests and to produce a set of practical guidelines for dealing with the results of these tests. The participants consisted of experts working in this area from academia, government, and industry. The various backgrounds and points of view represented helped to make the workshop productive in realistic terms.

Critical reviews of the "state-of-the-art" of various promising assays were presented by key scientists working in the field. Each has summarized his own remarks in Section I.

Prior to the meeting, CIIT drew up a proposed set of practical guidelines for using and interpreting the results of the short-term tests. These were discussed and modified by the workshop participants and are presented in Section II. A summary of the discussion and consensus of the group has been included following each major point. An effort was also made to incorporate dissenting views in order to put these issues in perspective.

Several hypothetical test cases were discussed and are included in Section III. The summary of each discussion leader is presented along with substantial portions of the actual transcript. These discussions are excellent examples of rational ways to approach these complex problems.

The opinions expressed by the Workshop participants are solely their own. Their input is gratefully acknowledged. However, any positions set forth in the document remain those of CIIT. Participation in the workshop does not necessarily mean support of these positions.

THE EDITOR-IN-CHIEF

Leon Golberg, M. B., Sc. D., Ph. D., FRIC, FRCPath., is President of the Chemical Industry Institute of Toxicology (CIIT), a research organization supported by 32 chemical companies with temporary headquarters in Raleigh, North Carolina. (Permanent headquarters in Research Triangle Park, N. C. are due to be completed in March, 1979.)

Dr. Golberg came to CIIT in February, 1976 from Albany Medical College, where he was Research Professor of Pathology and Scientific Director of the Institute of Comparative and Human Toxicology. Prior to joining Albany Medical College (in 1967), he was Director of the British Industrial Biological Research Association for six years. He also has experience in industry, having served as Medical Research Director of Benger Laboratories, Ltd. (now Fisons Pharmaceuticals, Ltd.).

Dr. Golberg received his Ph. D. in organic chemistry from the University of Oxford. He also holds a Sc. D. in biochemistry from Witwatersrand University, Johannesburg, and is a medical graduate of the University of Cambridge. He is a Fellow of the Royal Society of Medicine, the Royal Institute of Chemistry, and a Founder Fellow of the Royal College of Pathologists. He is President of the Society of Toxicology. Dr. Golberg, who has published extensively in the fields of synthetic organic chemistry, medicinal chemistry, biochemistry, metabolism, pharmacology, experimental pathology, and toxicology, is Editor-in-Chief of the CRC Toxicology Series.

THE EDITOR

Byron E. Butterworth, Ph. D., is Chief of Genetic Toxicology for the Chemical Industry Institute of Toxicology, Research Triangle Park, North Carolina.

He received his B.S. degree in chemistry in 1968 from Brigham Young University, Provo, Utah, and his Ph. D. degree in biochemistry from the University of Wisconsin, Madison, in 1972.

From 1972 to 1976, Dr. Butterworth was on the staff of the Central Research Department of E.I. du Pont de Nemours and Company, where he conducted research on the mechanism of action of antiviral agents. From 1976 to 1977, he was Chief of the Molecular Biology Section at du Pont's Haskell Laboratory for Toxicology and Industrial Medicine, where he directed the program for detecting potential mutagens/carcinogens utilizing short-term assays.

Dr. Butterworth is a member of the American Association for Cancer Research and the Environmental Mutagen Society. He is an Adjunct Assistant Professor in the Department of Pathology at the University of North Carolina in Chapel Hill. He has published many papers in the field of Molecular Biology.

PARTICIPANTS

Dr. Bernard D. Astill
Supervisor, Biochemistry
Health, Safety and Human Factors
 Laboratory
Eastman Kodak Co.
Rochester, New York

Dr. J. D. Behun
Research and Development
Mobil Chemical Co.
Edison, New Jersey

Dr. Z. G. Bell
PPG Industries, Inc.
Pittsburg, Pennsylvania

Dr. David J. Brusick
Director, Department of Genetics and
 Cell Biology
Litton Bionetics, Inc.
Kensington, Maryland

Dr. John A. Budny
Chief of Toxicology
Diamond Shamrock Corp.
Painesville, Ohio

Mr. John H. Butala
Information Toxicologist
Medical and Health Resources
 Division
Gulf Science and Technology Co.
Pittsburg, Pennsylvania

Dr. Byron E. Butterworth
Chief of Genetic Toxicology
Chemical Industry Institute of
 Toxicology
Research Triangle Park, North
 Carolina

Dr. E. C. Capaldi
The ARCO Chemical Co.
Glenolden, Pennsylvania

Dr. Daniel Casciano
Division of Mutagenesis Research
National Center for Toxicological
 Research
Jefferson, Arkansas

Dr. Donald Clive
Burroughs Wellcome
Research Triangle Park, North
 Carolina

Dr. Ralph R. Cook
Dow Chemical U.S.A.
Chapel Hill, North Carolina

Dr. David B. Couch
Genetic Toxicology Department
Chemical Industry Institute of
 Toxicology
Research Triangle Park, North
 Carolina

Dr. E. M. Dixon
Corporate Medical Director
Celanese Corp.
New York, New York

Dr. Virginia Dunkel
National Cancer Institute
Bethesda, Maryland

Dr. W. Gary Flamm
Food and Drug Administration
Branch Chief, Genetic Toxicology
Washington, D.C.

Dr. Marvin A. Friedman
Allied Chemical Corp.
Morristown, New Jersey

Dr. James E. Gibson
Vice President and Director of
 Research
Chemical Industry Institute of
 Toxicology
Research Triangle Park, North
 Carolina

Dr. Leon Golberg
President
Chemical Industry Institute of
 Toxicology
Research Triangle Park, North
 Carolina

Dr. G. Kent Hatfield
Toxicologist
Diamond Shamrock Corp.
Cleveland, Ohio

Dr. Richard Henderson
Director, Health Sciences and
  Toxicology Research
Olin Corp.
New Haven, Connecticut

Dr. Robert K. Hinderer
Senior Environmental Toxicologist
B. F. Goodrich Chemical Co.
Cleveland, Ohio

Dr. Mark Hite
Director of Toxicology and Pathology
Merck Institute for Therapeutic
  Research
West Point, Pennsylvania

Dr. Abraham W. Hsie
Biology Division
Oak Ridge National Laboratory
Oak Ridge, Tennessee

Dr. Don Hughes
Miami Valley Laboratories
Procter & Gamble Co.
Cincinnati, Ohio

Dr. Gary Johnson
Miami Valley Laboratories
Proctor & Gamble Co.
Cincinati, Ohio

Dr. D. J. Kilian
Regional Director, Occupational
  Health and Medical Research
Dow Chemical U.S.A.
Freeport, Texas

Mr. Carroll J. Kirwin
Industrial Toxicologist
Medical Division
Phillips Chemical Co.
Bartlesville, Oklahoma

Dr. David F. Krahn
Chief, Molecular Biology Section
Haskell Laboratory for Toxicology
  and Industrial Medicine
E. I. du Pont de Nemours & Co., Inc.
Newark, Delaware

Dr. Robert Kuna
Research and Environmental Health
  Division
Exxon Corp.
Linden, New Jersey

Dr. Dean Lampe
Northern Natural Gas
Omaha, Nebraska

Dr. Marvin S. Legator
Professor and Director
Department of Preventive Medicine
  and Community Health
Division of Environmental Toxicology
The University of Texas Medical
  Branch
Galveston, Texas

Dr. George J. Levinskas
Manager, Environmental Assessment
  and Toxicology
Monsanto Co.
St. Louis, Missouri

Dr. Eric Longstaff
Imperial Chemical Industries, Ltd.
Central Toxicology Laboratory
Cheshire, England

Dr. J. R. Lovett
Vice President — Research
Air Products and Chemicals, Inc.
Allentown, Pennsylvania

Mr. R. W. McKinney
W. R. Grace & Co.
Columbia, Maryland

Dr. Myron A. Mehlman
Director, Environmental Health and
  Toxicology
Mobil Oil Corp.
New York, New York

Dr. Paul O. Nees
Corporate Toxicologist
Hooker Chemicals and Plastics Corp.
Niagara Falls, New York

Dr. Roman J. Pienta
Deputy Director
Chemical Carcinogenesis Program
Head, *In Vitro* Carcinogenesis
Frederick Cancer Research Center
Frederick, Maryland

Mr. J. M. Ramey
Manager, Product Standards
Celanese Corp.
New York, New York

Dr. Verne A. Ray
Assistant Director
Safety Evaluation Department
Medical Research Laboratories
Pfizer, Inc.
Groton, Connecticut

Mr. Ben C. Rusche
Administration Department
E. I. du Pont de Nemours & Co., Inc.
Wilmington, Delaware

Dr. M. W. Sauerhoff
Stauffer Chemical Co.
Westport, Connecticut

Dr. Gerald P. Schoenig
Manager of Toxicology
Agricultural Chemical Group
FMC Corp.
Middleport, New York

Dr. Harvey E. Scribner
Toxicology Department
Rohm and Haas Co.
Spring House, Pennsylvania

Dr. Bernard A. Schwetz
Toxicology Research Laboratory
Dow Chemical U.S.A.
Midland, Michigan

Dr. William Sheridan
Mammalian Genetics Section
Laboratory of Environmental
  Mutagenesis
National Institute of Environmental
  Health Sciences
Research Triangle Park, North
  Carolina

Dr. R. S. Slesinski
Carnegie-Mellon Institute of Research
Pittsburgh, Pennsylvania

Dr. M. B. Slomka
Consulting Toxicologist
Health, Safety, and Environment
Shell Oil Co.
Houston, Texas

Dr. T. H. F. Smith
Manager of Toxicology and Product
  Safety
Tenneco Chemicals, Inc.
Saddle Brook, New Jersey

Dr. W. M. Smith
Group R&D Coordinator
Air Products and Chemicals, Inc.
Allentown, Pennsylvania

Dr. Eugene R. Soares
Genetic Toxicology Department
Chemical Industry Institute of
  Toxicology
Research Triangle Park, North
  Carolina

Dr. James A. Swenberg
Chief of Pathology
Chemical Industry Institute of
 Toxicology
Research Triangle Park, North
 Carolina

Dr. William Thilly
Department of Nutrition and Food
 Science
Massachusetts Institute of Technology
Cambridge, Massachusetts

Ms. Judy Tins
Manager, Toxicological and Product
 Safety
Celanese Corp.
New York, New York

Dr. Gary M. Williams
Chief, Division of Experimental Pathology
American Health Foundation
Naylor Dana Institute for Disease
 Prevention
Valhalla, New York

CONTRIBUTORS

Patricia A. Brimer, B.S.
Biological Laboratory Specialist
Oak Ridge National Laboratory
Oak Ridge, Tennessee

David J. Brusick, Ph. D.
Director
Department of Genetics and Cell
 Biology
Linton Bionetics, Inc.
Kensington, Maryland

Byron E. Butterworth, Ph.D.
Chief of Genetic Toxicology
Chemical Industry Institute of
 Toxicology
Research Triangle Park, North
 Carolina

Donald Clive, Ph.D.
Research Scientist and Head
Genetic Toxicology Laboratory
Burroughs Wellcome Co.
Research Triangle Park, North
 Carolina

David B. Couch, Ph.D.
Staff Scientist
Chemical Industry Institute of
 Toxicology
Research Triangle Park, North
 Carolina

Virginia C. Dunkel, Ph.D.
Coordinator
In vitro Carcinogenesis Program
Division of Cancer Cause and
 Prevention
National Cancer Institute
Bethesda, Maryland

Nancy L. Forbes
Biological Laboratory Technician
Oak Ridge National Laboratory
Oak Ridge, Tennessee

James C. Fuscoe, B.S.
Graduate Student in Biomedical
 Science
University of Tennessee
Oak Ridge, Tennessee

Mark Hite, Sc.D.
Director of Toxicology and Pathology
Merck Institute for Therapeutic
 Research
West Point, Pennsylvania

Abraham W. Hsie, Ph.D.
Senior Research Staff Scientist
Biology Division
Oak Ridge National Laboratory
Oak Ridge, Tennessee

Mayphoon H. Hsie, B.S.
Consulting Statistician
Biology Division
Oak Ridge National Laboratory
Oak Ridge, Tennessee

David F. Krahn, Ph.D.
Chief, Molecular Biology Section
Haskell Laboratory for Toxicology
 and Industrial Medicine
E. I. du Pont de Nemours & Co., Inc.
Newark, Delaware

Robert Kuna, Ph.D.
Toxicologist
Research and Environmental Health
 Division
Exxon Corporation
Linden, New Jersey

Marvin S. Legator, Ph.D.
Director
Division of Environmental Toxicology
The University of Texas Medical
 Branch
Galveston, Texas

Albert P. Li, Ph.D.
Assistant Professor
Cancer Research and Treatment Center
University of New Mexico
Albuquerque, New Mexico

Howard L. Liber
NIEHS Predoctral Trainee
Massachusetts Institute of Technology
Cambridge, Massachusetts

Eric Longstaff, Ph.D.
Acting Section Head
Experimental Pathology Section
Imperial Chemical Industries, Ltd.
Central Toxicology Laboratory
Cheshire, England

Richard Machanoff, B.S.
Biological Laboratory Technician
Oak Ridge National Laboratory
Oak Ridge, Tennessee

J. Patrick O'Neill, Ph.D.
Research Staff Scientist
Biology Division
Oak Ridge National Laboratory
Oak Ridge, Tennessee

Gary L. Petzold, M.S.
Biology and Biochemistry Assistant
Pathology and Toxicology Unit
The Upjohn Co.
Kalamazoo, Michigan

Roman J. Pienta, Ph.D.
Deputy Director
Chemical Carcinogenesis Program
Frederick Cancer Research Center
Frederick, Maryland

Verne A. Ray, Ph.D.
Assistant Director
Safety Evaluation Department
Medical Research Laboratories
Pfizer, Inc.
Groton, Connecticut

James C. Riddle, Ph.D.
Biomedical Sciences
University of Tennessee
Oak Ridge, Tennessee

Juan R. San Sebastion, Ph.D.
Postdoctoral Investigator
Biology Division
Oak Ridge National Laboratory
Oak Ridge, Tennessee

Bernard A. Schwetz, D.V.M.
Research Manager
Toxicology Research Laboratory
Dow Chemical U.S.A.
Midland, Michigan

William Sheridan, Sc.D.
Research Geneticist
Laboratory of Environmental
  Mutagenesis
National Institute of Environmental
  Health Sciences
Research Triangle Park, North
  Carolina

Eugene R. Soares, Ph.D.
Mammalian Geneticist
Genetic Toxicology Department
Chemical Industry Institute of
  Toxicology
Research Triangle Park, North
  Carolina

James A. Swenberg, Ph.D.
Chief of Pathology
Chemical Industry Institute of
  Toxicology
Research Triangle Park, North
  Carolina

William G. Thilly, Sc.D.
Associate Professor of Genetic
  Toxicology
Massachusetts Institute of Technology
Cambridge, Massachusetts

Gary M. Williams, M.D.
Chief, Division of Experimental
  Pathology
American Health Foundation
Naylor Dana Institute for Disease
  Prevention
Valhalla, New York

ACKNOWLEDGMENT

I would like to thank Dr. Leon Golberg for his substantial contribution in editing this manuscript and help with the arduous task of transforming transcripts of conversations into a readable form. Any errors that remain are, of course, my own.

B.E.B.

TABLE OF CONTENTS

I. CRITICAL REVIEWS OF THE STATE OF THE ART FOR SHORT-TERM TESTS

Observations and Recommendations Regarding Routine Use of Bacterial Mutagenesis Assays as Indicators of Potential Chemical Carcinogens 3
David J. Brusick

A Discussion of Gene Locus Mutation Assays in Bacterial, Rodent, and Human Cells 13
William G. Thilly and Howard L. Liber

Utilization of a Quantitative Mammalian Cell Mutation System, CHO/HGPRT, in Experimental Mutagenesis and Genetic Toxicology 39
Abraham W. Hsie, David B. Couch, J. Patrick O'Neill, Juan R. San Sebastion, Patricia A. Brimer, Richard Machanoff, James C. Riddle, Albert P. Li, James C. Fuscoe, Nancy Forbes, and Mayphoon H. Hsie

Cultured Mammalian Cell Transformation Systems 55
David F. Krahn

Genetic Aspects of Short-term Testing and an Appraisal of Some In Vivo Mammalian Test Systems 67
Eugene R. Soares

The Usefulness of DNA Damage and Repair Assays for Predicting Carcinogenic Potential of Chemicals 77
James A. Swenberg and Gary L. Petzold

II. PRACTICAL GUIDELINES FOR THE USE AND INTERPRETATION OF SHORT-TERM TESTING FOR MUTAGENS/CARCINOGENS

Recommendations for Practical Strategies for Short-Term Testing for Mutagens/Carcinogens 89
Byron E. Butterworth, Eric Longstaff, Donald Clive, Roman J. Pienta, Robert Kuna, Marvin S. Legator, William G. Thilly, Verne A. Ray, William Sheridan, Gary M. Williams, and Virginia Dunkel

III. DISCUSSION OF POSSIBLE COURSES OF ACTION GIVEN SEVERAL HYPOTHETICAL TEST CASES

Discussion of Hypothetical Test Case A 105
Bernard A. Schwetz

Discussion of Hypothetical Test Case B 117
Verne A. Ray

Discussion of Hypothetical Test Case C 123
Mark Hite

Index 131

Section I
Critical Reviews of the State of the Art for Short-Term Tests

OBSERVATIONS AND RECOMMENDATIONS REGARDING ROUTINE USE OF BACTERIAL MUTAGENESIS ASSAYS AS INDICATORS OF POTENTIAL CHEMICAL CARCINOGENS

D. J. Brusick

TABLE OF CONTENTS

I. Introduction 3

II. Assay System Variability 8
    A. Strain Quality Control 8
    B. Activation Quality Control 9

III. Compound Variability 10

IV. Variability in the Interpretation of Tests Results 10

References 12

I. INTRODUCTION

Bacterial assays capable of detecting genetic activity of chemicals or their breakdown products come in many forms and utilize many different species and strains of bacteria (see Table 1 for references). The protocols, also varied, can be classified into a few general types and these are given in Figures 1 through 6.

Bacterial tests of the type outlined in the figures are used increasingly in screening chemicals for genetic activity. The use of these tests has spread from the research laboratory into the toxicology and safety-testing arena, and the results of bacterial tests, such as those described by Ames and co-workers,[1-3] are reported frequently in the news media, often with far-reaching implications attached. This paper presents an attempt to define the advantages and limitations of bacterial tests and to put forth some general considerations regarding the use of these assays in routine, chemical safety evaluations.

The following observations and recommendations are directed toward the use of bacterial mutation systems as routine assays for determining the biological properties of nonclassified ("unknown") agents. The problems that arise in this application are often unique and almost never the same as the problems encoun-

TABLE 1

Microbial Assay Systems for Mutagens/Carcinogens

Test system	Ref.
Salmonella typhimurium (Ames)	9, 10
S. typhimurium (host-mediated)	11
Escherichia coli (Bridges)	12
E. coli (T4 bacteriophage)	13
E. coli (prophage induction)	14
Bacillus subtilis (transforming DNA)	15
B. subtilis (inhibition)	16
B. subtilis (spores)	17
Klebsiella pneumoniae	18
Citrobacter freundii	18
Dictyostelium discoideum	19
Streptomyces coelicolor	20
Saccharomyces cerevisiae	21, 22
S. pombe	23
Aspergillus nidulans	24
Neurospora crassa	25

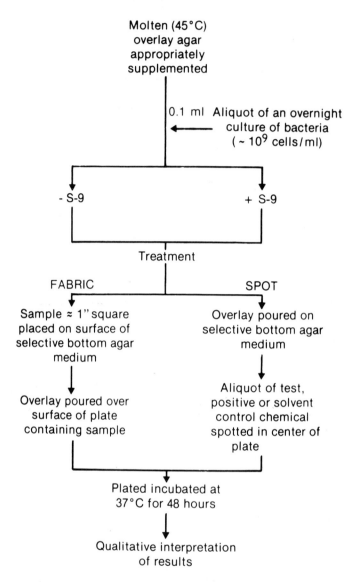

FIGURE 1. Flow chart for the spot test procedure using chemical substances or fabrics. This assay depends on high solubility and diffusion of the test material. The results are qualitative in nature and can be used as a quick screen.

tered in research laboratories, where one or a small number of classical mutagens are intensely investigated.

Industrial application of this type of testing generally necessitates that scientific and management decisions be made on the basis of results from these tests; "these tests" refer in most instances to the Ames *Salmonella*/microsome assay.[3] The following problems must be dealt with in bacterial testing:

1. What constitutes a positive or negative finding? Since the chemicals being evaluated are undesignated, the development of a set of criteria upon which a + or − designation is made becomes critical. To es-

REVERSE MUTATION ASSAY
[Agar Incorporation Method]

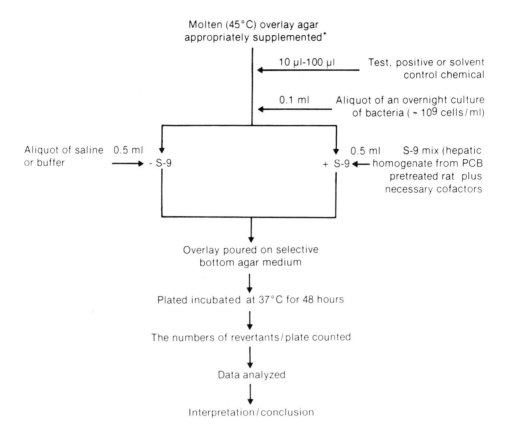

FIGURE 2. Flow chart for the agar incorporation method or "standard" Ames *Salmonella*/microsome assay. Chemicals of low solubility and diffusion can be evaluated in this assay. The results are semiquantitative in nature, and a dose-response effect is usually part of the criteria used to identify a mutagenic material.

tablish such criteria, the inherent resolving power of the assay system must be determined. Without uniform criteria the same data configuration may be designated positive by some laboratories and negative by others. This results in confusion and may lead to an opinion that the tests are giving nonreproducible and thus useless results. The nonreproducibility in these situations, however, rests with variability in interpretation of negative and positive effects and not with the assay itself.

2. If a clearly positive set of data is obtained, how can the tester be confident that the effect is compound-induced and not an artifact of the test system or the methodology employed? The question of developing a standardized protocol is often raised and quickly "shelved" using the rationale that it is premature to develop a fixed protocol for such new tests. While this may be the case, it is not premature, in my opinion, to *define* a generally acceptable protocol for use in routine screening. The definition of

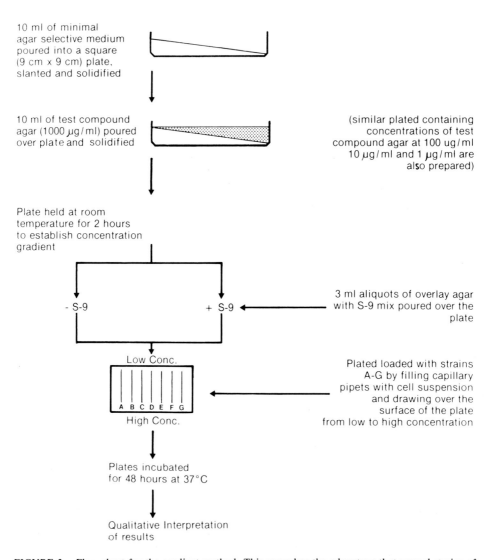

FIGURE 3. Flow chart for the gradient method. This assay has the advantage that several strains of bacteria can be used to screen a single compound incorporated into a set of two or three plates. This test depends on solubility and diffusion. The results are strictly qualitative in nature. Similar to other plate assays, only strains measuring back mutation can be employed.

a generally recognized protocol would not negate results from variations or modifications of the recognized protocol, but it would provide a guide for laboratories desiring a uniform assay for general screening. The protocol would be designed to ensure reasonable success at detecting all positive chemicals.

3. How should the results of bacterial tests be used in making decisions regarding potential hazard? It has been proposed that positive results in an Ames test constitute the identification of a probable carcinogen and that immediate notification be given to the appropriate regulatory agency and potentially exposed populations (i. e., exposed workers). On the other hand, if the Ames test is part of a tier or multiphasic ap-

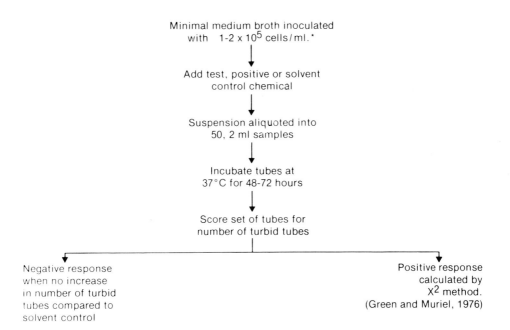

FIGURE 4. Flow chart for the modified fluctuation method. This procedure might be advantagous when a "standard" Ames test gives equivocal results since it uses a statistical approach to evaluation of the data. Conditions for use of S9 mix might be suboptimal with this technique.

proach to chemical screening, it might be advantagous to wait until all the short-term testing is complete before taking action. Action taken on the basis of a single-test result may be premature and cause undue confusion if the test results are not substantiated by other test systems. A recent regulatory decision involving such a situation has indicated that negative results from higher order, short-term tests may not "erase" the carcinogen indictment of an Ames test.[4]

4. When a chemical does not produce a mutagenic response in a bacteria assay, does this constitute a reliable demonstration of nonmutagenicity? It is important for the further investment of resources into the development of a chemical to be confident that the test which has been conducted constitutes a valid attempt to establish the genetic properties of the chemical in question. The utility of the bacterial test systems would be significantly reduced if it were established that a high proportion of negative compounds produced positive responses in higher order tests. Already certain types or classes of chemicals have been identified as "poor responders" in the Ames test, and negative results with derivatives or structurally related analogues of these chemicals should be reviewed with suspicion.

Thorough discussion and resolution of these general questions must be undertaken before bacterial assays can acquire regulatory status. The answers will not come easily and a great deal of scientific input will be required. I would like to review some of the steps we have taken in our laboratory to provide a workable screening program with bacterial assays. Most of the

DNA MODIFICATION ASSAYS

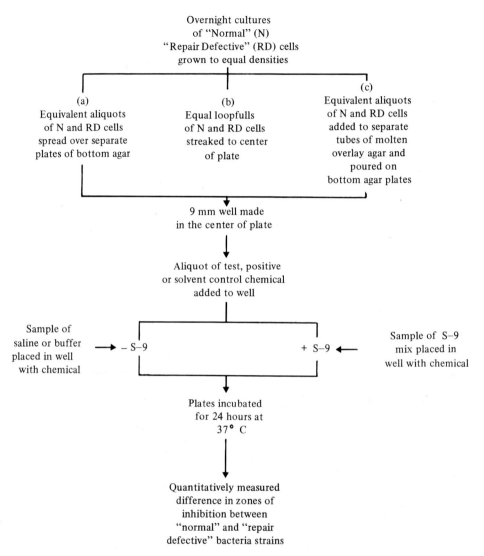

FIGURE 5. Flow chart for various DNA repair tests. These assays do not measure mutational events, but rather measure the extent of repairable DNA damage induced by a chemical. Equal levels of toxicity in N and RD cells indicate no DNA type cytotoxicity, whereas greater toxicity in RD cells than in N cells indicates a chemical capable of DNA damage. The results are qualitative in nature.

procedures have been based on several years of experience, involving research and actual testing applications.

We have identified three areas of potential variability in conducting bacterial tests.

II. ASSAY SYSTEM VARIABILITY

Standardization of all test system components is necessary to ensure maximum sensitivity and reproducibility. These steps will not impose a fixed protocol but will define acceptable standards under which any protocol can be used with the confidence that the results will be valid and not artificial. The following steps are recommended to minimize assay system variability.

A. Strain Quality Control

Technical verification of strain identity

HOST-MEDIATED ASSAY SCHEME

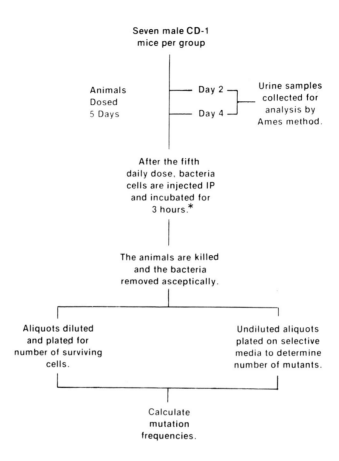

FIGURE 6. Flow chart for host-mediated type tests. This method employs the whole animal as the source of metabolism. Both urine from treated animals and cells recovered from the host are assayed. Sensitivity of the method depends on the systematic distribution and/or excretion parameters of the administered compound and its metabolites.

should be a part of all reports, publications, and protocol-support documents. This should include storage and culture methods, as well as the results of tests to monitor all relevant mutant "marker" genes. Evidence of noncontamination and induced reversion by reference mutagens should also be included. Establishment of a reference laboratory where strain identity can be verified would be highly desirable. The concern for strain-identity results from the possibility of strain drift (subtle changes in genetic background following subculture which alter the growth and physiological properties of microbial cultures). Strain drift can be reduced by the preparation of frozen cultures in DMSO according to Ames.[3] Working stocks may shift slightly and the preparation of new frozen stocks might results in strain drift over a period of years.

B. Activation Quality Control

Activation quality control should apply

equally to the preparation, storage, and utilization of tissue homogenate S9 fractions. S9 preparations should be made in lots with batch identification. The animal source and strain from which tissues are obtained should be identified on permanent records. Certain biochemical and biological parameters of an S9 preparation should be measured for each batch and recorded. Although there is no current method of S9 preparation that can guarantee the preservation of all in vivo enzymatic activity, measurements of the milligrams protein per milliliter, relative P450/P448 activity (i. e., direct measurement of arylhydrocarbon hydroxylase [AHH] activity) and activation of reference promutagens can provide a basis for comparison of batches. Identification of other test-system components, such as solvents and media, should be maintained when possible.

Without adequate quality control of its components, the assay system cannot be relied upon to provide reliable and reproducible results. Identifying quality control measures in publications and other report documents will increase the level of confidence in the reliability of the results and their interpretation.

III. COMPOUND VARIABILITY

Most liquids and solid materials with reasonable solubility in water or DMSO can be evaluated over a wide series of dose levels. Visible compound insolubility in the test system should not be used to determine the maximum test level. Good dose-response curves for mutagenicity can be achieved above compound concentrations which produce visible precipitation. The best method to determine the high dose level is a weak to moderate indication of toxicity, usually evident by a slight decline in the number of spontaneous mutants per plate. The lowest concentration in the series should be in the clearly nontoxic range.

There are substances, however, that are not easily subjected to these criteria. They include gaseous and highly volatile materials; "completely insoluble" materials such as fabrics, paper, fibers, and some plastic polymers; and complex mixtures of chemicals.

Gaseous and volatile materials can be evaluated through the use of various types of closed-chamber systems. Dessicator jars, anaerobic chambers, and dessicator cabinets have been used successfully to measure the mutagenicity of these substances. Protocols in which a fixed volume (concentration) of gas or volatile agent have been introduced into the chamber and the exposure time varied (usually 1, 6, 24, and 48 hr) and protocols in which varied concentrations of the gas or volatile agent are introduced and incubated for a fixed interval (usually 48 hr) have been employed. We have had best results with a protocol that involves the introduction of a fixed volume of material into a chamber with exposure times of 1, 6, 12, 24, and 48 hr followed by additional postexposure incubation of all test and control plates.[8] In a limited survey of chemicals, certain physical properties, especially water solubility, were important in generating a positive response. Much more research into gas and vapor exposure needs to be conducted before these compounds can be routinely screened.

The preponderance of published data using the microbial systems involves single substances. Evaluation of mixtures, such as environmental air and water samples, petroleum fractions, and industrial waste effluents, can generate some difficult problems. The most perplexing has been the mixture that contains a minor concentration of mutagen and a major component that is toxic but nonmutagenic. This type of mixture will often result in negative test results because the test system is killed by the major nonmutagenic component before the level of the mutagen is sufficient to produce clear activity. The resolution of this problem is extremely difficult, unless the mixture can be fractionated and each subfraction evaluated. With complex mixtures this would generally not be possible.

IV. VARIABILITY IN THE INTERPRETATION OF TEST RESULTS

The need for criteria development was identified earlier. The characteristics of a good criteria package should be that in the screening of unknown agents a positive effect can be identified with a 95% confidence level and at the same time the number of "false" positives resulting from random fluctuations in the number of spontaneous mutants per plate should be almost zero. Application of statistical analysis and the use of a high number of replicate plates have been suggested as possible parameters to eliminate evaluation errors. While these may have some benefit, the nature of the bacterial assay as described by Ames does not lend itself easily to valid statistical analysis. Multiple replicates have the limitation that errors in test system preparation or chemical dilutions, etc. will simply be replicated.

An approach that appears to have the greatest application consists of defining the limits of the spontaneous background and then outlining the effects required for designating a compound to be active. As a result of experience with almost 2000 compounds we have selected the following set of criteria for evaluation of Ames test data:

1. A defined range for spontaneous fluctuation has been developed over a 2-year period, under conditions of routine testing, for bacterial strains TA-1535, TA-1537, TA-1538, TA-98, and TA-100. If a set of control negative or solvent control data for a given test falls outside that range, the test run for that strain is rejected and repeated.
2. The dose range used is wide, with doses separated by log or half-log steps. The highest concentration must show some signs of toxicity or reach a level of 2000 μg per plate without toxicity. The low of the concentration range must be in a clearly nontoxic area.
3. The criteria for a positive effect for strains TA-1535, TA-1537, and TA-1538 are that a dose-response effect must be evident over at least three concentrations separated by at least half-log steps and that the minimum level of response in the dose-related effect be 2.0 times the concurrent solvent control value. For strains TA-98 and TA-100, a dose-related response must also be observed over three concentrations separated by at least half-log steps, and the maximum level of response in the dose-related effect must be 2.0 times the concurrent solvent control value. The larger spontaneous frequency for TA-98 and TA-100 makes smaller increases more significant.
4. Any points that show the appropriate 2.0 times background but are not within a dose-related trend are often repeated with a narrower dose-range cluster above and below that which showed a response.

Employing these test evaluation criteria will provide a sound guide for the accurate identification of chemicals mutagenic in the Ames *Salmonella*/microsome assay, as well as related bacterial reverse-mutation assays. The application of such criteria will also provide chronologically consistent evaluations to the test data. This is important when criteria are used in making management or regulatory decisions. To apply one set of criteria this year and a different set next year makes comparative assessments impossible. Before any formal statistical approach is adopted for plate test-data evaluation it should be thoroughly tested on a wide array of published data. Because of the ease of test reproduction, the use of chronological reproducibility may be a more powerful tool than a statistical analysis of a single run in the verification of suspicious data. The use of a modified fluctuation test, as described by Green and Muriel,[5] is another possible approach to accurately classify effects that are outside the resolving power of the standard Ames approach.

The development of evaluation criteria for other types of plate assays such as the *Escherichia coli pol A+/pol A*/DNA repair assay[7] and the "rec" assay with *Bacillus*[6] will also be important as these tests become more widely applied to chemical screening.

It must be noted at this point that even the best set of criteria cannot be applied with absolute confidence. Sound scientific judgment on the part of the investigators is an important ingredient in the overall test evaluation. The bacterial reverse-mutation tests are deceptively simple, while in fact the biological and chemical reactions that occur in the overlay agar containing the chemical, cells, solvent, and S9 mix are hardly understood at all. Tests such as the Ames *Salmonella*/microsome assays have the potential to be an extremely powerful tool in the overall assessment of biological properties of chemicals. Misuse in making rapid judgments of human risk from results of these tests is tempting because of their apparent simplicity and good correlation with other toxicological endpoints.

There is an increasing tendency to use the Ames test to substantiate human harm or safety without the development of a complete data profile on the chemical in question. Unless the scientific community acts promptly to answer some of the questions stated previously and to define the proper testing conditions for use of the bacterial assays, nonscientific misuse of these tests may result in the premature loss of these potentially valuable toxicological tools.

REFERENCES

1. **Ames, B. N.**, The detection of chemical mutagens with enteric bacteria, in *Chemical Mutagens: Principles and Methods for Their Detection*, Vol. 1, Hollaender, E. A., Ed., Plenum Press, New York, 1971, 267.
2. **Ames, B. N., Durston, W. E., Yamasaki, E., and Lee, F. D.**, Carcinogens are mutagens: a simple test system combining liver homogenates for activation and bacteria for detection. *Proc. Natl. Acad. Sci. U. S. A.*, 70, 2281, 1973.
3. **Ames, B. N., McCann, J., and Yamasaki, E.**, Methods for detecting carcinogens and mutagens with the *Salmonella*/mammalian-microsome mutagenicity test, *Mutat. Res.*, 31, 347, 1975.
4. Administrative Decision in the Matter of Acrylonitrile Copolymers Used to Fabricate Beverage Containers, Dockett No. 76N-0070, Judge Davidson, U. S. Food and Drug Administration, August 1977.
5. **Green, M. H. L. and Muriel, W. S.**, Mutagen testing using trp^+ reversion in *Escherichia coli*, *Mutat. Res.*, 38, 3, 1976.
6. **Kada, T., Tutikawa, K., and Sadaie, Y.**, *In vitro* and host-mediated recassay procedures for screening chemical mutagens and phloxine, a mutagenic red dye detected, *Mutat. Res.*, 16, 165, 1972.
7. **Slater, E. E., Anderson, M. D., and Rosenkranz, H. S.**, Rapid detection of mutagens and carcinogens, *Cancer Res.*, 31, 970, 1971.
8. **Jagannath, D. R., Bakshi, K., and Brusick, D. J.**, in preparation.
9. **Ames, B. N., Lee, F. D., and Durston, W. E.**, *Proc. Natl. Acad. Sci. U. S. A.*, 70, 782, 1973.
10. **McCann, J., et al.**, *Proc. Natl. Acad. Sci. U. S. A.*, 72, 5125, 1975.
11. **Legator, M. S. and Malling, H. V.**, *Chemical Mutagens: Principle Methods for Their Detection*, Vol. 2, Hollaender, A., Ed., Plenum Press, New York, 1971, 586.
12. **Bridges, B. A.**, *Lab. Pract.*, 21, 413, 1972.
13. **Hartman, P. E. et al.**, *Science*, 172, 1058, 1971.
14. **Goldschmidt, F. P., Miller, R., and Matney, S. T.**, *Microbiol. Genet. Bull.*, 41, 3, 1976.
15. **Freese, E. and Strack, H. B.**, *Proc. Natl. Acad. Sci. U. S. A.*, 48, 1796, 1962.
16. **Kada, T.**, *Mutat. Res.*, 38, 271, 1976.
17. **MacGregor, J. T. and Sacks, L. E.**, *Mutat. Res.*, 38, 271, 1976.
18. **Kramers, P. G. N., Knapp, A. G. A. C., and Voogd, C. E.**, *Mutat. Res.*, 31, 65, 1975.
19. **Liwerant, I. J. and Pereira Da Silva, L. H.**, *Mutat. Res.*, 33, 135, 1976.
20. **Carere, A. et al.**, *Mutat. Res.*, 38, 136, 1976.
21. **Parry, J. M.**, *Lab. Pract.*, 21, 417, 1972.
22. **Chambers, C. and Dutla, S. K.**, *Genetics*, 83 (Suppl.), 13, 1976.
23. **Loprieno, N.**, *Mutat. Res.*, 29, 237, 1975.
24. **Kaffer, E., Marshall, P., and Cohen, G.**, *Mutat. Res.*, 38, 141, 1976.
25. **de Serres, F. J.**, *Chemical Mutagens: Principle Methods for their Detection*, Vol. 2, Hollaender, A., Ed., Plenum Press, New York, 1971, 311.

A DISCUSSION OF GENE LOCUS MUTATION ASSAYS IN BACTERIAL, RODENT, AND HUMAN CELLS

W. G. Thilly and H. L. Liber

TABLE OF CONTENTS

I. Introduction 13

II. Discussion 13
    A. Mammalian Cell Mutation Assays 13
    B. Bacterial Gene Locus Mutation Assay 19
    C. Drug Metabolism in Mutation Assays 26
    D. Gene Structure Affects the Probability of Observing Mutation 26

III. Conclusions 36

References 36

I. INTRODUCTION

Responsible control of chemicals in the environment requires reliable and effective means of differentiating hazardous from nonhazardous materials. Since our first responsibility is the protection of humans from chemically induced cancer and genetic disease, it is important to recognize the requirement that bioassay systems faithfully reflect the biochemical specificity of human tissues responding to chemical exposure.

The high cost of long-term animal tests has led to the development and application of faster, cheaper assays of genetic damage, utilizing bacteria, fungi, or mammalian cells in culture. The known differences between man and rodents or microbial species, with respect to drug metabolism and DNA repair systems, suggest the need for the development and use of mutation and cell transformation assays, with diploid human cells coupled to a drug-metabolizing system derived from human liver. The current scientific literature does much to confirm our conception of the need for such systems, both for bioassay purposes and as experimental models for the study of mutagenesis and oncogenesis. Our laboratory at MIT has developed a reliable quantitative assay for mutagenic activity in diploid human lymphoblasts. A potentially useful addition to the battery of bacterial mutation assays has also been developed. A rat-liver-derived drug-metabolizing element, active for a broad range of chemical classes, easily prepared sterilely, and compatible with the human cell, is used in our mutation assays.

In these few pages a review of the crucial issues of gene locus mutation assays has been attempted; first examining the various mammalian cell mutation assays in use today and then turning attention to bacterial mutation systems. Data are presented comparing the sensitivity of forward and reverse assays in *Salmonella typhimurium*. The review closes with a discussion of how gene structure affects observable mutation and thus suggests a preference for a forward mutation system.

II. DISCUSSION

A. Mammalian Cell Mutation Assays

While mutation assays utilizing prokaryotes and various eukaryotes, such as yeast and other fungi, have been in use for three decades,[30,60] the development of similar systems employing mammalian cells in culture is a relatively recent event.[22,61]

The first two lines to be successfully mutated in vitro were the V79 and CHO lines, both of Chinese hamster origin. CHO was derived from normal ovarian tissue and V79 from normal lung. While the original studies utilized auxotrophy for glycine as the genetic marker,[61,62] most studies have involved selection for resistance to the purine analogs 8-azaguanine (8AG) or 6-thioguanine (6TG),[1,9,15,21,46,47,52,93] a trait generally agreed to be conferred by inactivation or deletion of the enzyme hypoxanthine-guanine phosphoribolsyltransferase. The reasons for the extensive use of this system are manifold. First, the sex-linked character of this enzyme allows utilization of normal diploid lines. Second, it is quite general with regard to the classes of mutagens to which it will respond, the only criterion being enzyme inactivation. Finally, a knowledge of the precise mutabolic lesion allows characterization of mutants by various immunochemical (cross-reacting material) and biochemical (complementation, enzyme assay, peptide sequencing) techniques.[11,57]

In recent years, a number of other markers have been exploited in these same cell lines. These include cyclohexamide resistance,[104] ouabain resistance, and reversion from a temperature-sensitive lesion involving protein synthesis.[52] A theoretical caution must be added because they should, in general, be insensitive to a wide variety of mutagens due to the highly specific nature of the lesion necessary to produce a response (see Section II.B). While ouabain resistance involves forward mutation, it suffers from molecular specificity, i.e., it requires that a region on a Na^+/K^+ATPase be altered, so that ouabain-binding ability is lost, while ATPase activity remains substantially unchanged. One would, therefore, expect this marker to behave similarly to a reversion-type assay. The low mutant frequencies obtained in these systems (10^{-7} to 10^{-5}) do in fact bear out these expectations.[52]

In addition to CHO and V79, several other cell lines have been used with various markers. To date, the most widely used is the L5178Y mouse lymphoma line. Clive[23] used a thymidine kinase (TK) heterozygote variant of this line (TK^+/TK^-) and assayed for resistance to bromodeoxyuridine, a trait conferred by the inactivation of thymidine kinase in much the same way that inactivation of HGPRT confers 8AG and 6TG resistance. Fox and Anderson[34] have examined the characteristics of mutants resistant to excess thymidine and iododeoxyuridine in both L5178Y and P388 mouse lymphoma cells and concluded that, unlike L5178Y, P388 is already heterozygous at the TK locus.

Another marker which has been used in the L5178Y line is phenotypic reversion from asparagine auxotrophy to prototrophy.[90] The genetic basis of this marker is obscure. Low mutant fractions (10^{-6} to 5×10^{-5}) seem to be indicative of a true genetic reversion, but the expected differential response between base-pair substitution mutagens and frameshift mutagens was not noted by Summers[90] or by Capizzi,[18] leading them to postulate a forward mutation leading to loss of a control protein. Morrow et al.[72] have obtained conflicting data, however, and the question remains unresolved. Finally, Knapp and Simons[63] have reported the successful use of the $6TG^R$ marker in this same cell line.

A number of investigators have studied human fibroblast strains of limited life span as target organisms in mutagenesis studies.[2,25,67] A novel marker, the ability to utilize fructose as a sole carbon source, was used in one of these studies.[25] The genetic basis of this selection system is obscure, however. DeMars'[2] and Maher's[67] laboratories have assayed for mutation to 8AG resistance.

Only one report of mutagenesis in an established human cell line appeared in the literature[81] before the recent publication of reports from our laboratory. Our studies, however, raise serious questions about the interpretation of this earlier report. Chemical mutagenesis in diploid human fibroblasts has been described recently.[59]

Table 1 is an attempt to summarize the results of a comprehensive literature survey of quantitative mutation determinations using mammalian cells. The principal steps in the mutagenesis assays summarized in Table 1 are essentially identical. First, cells are exposed to a known concentration of test compound for a fixed period of time. Second, *growth postexposure* under nonselective conditions occurs to

TABLE 1

In Vitro Mammalian Mutagenesis

Hamster CHO Cells

Agent	Dose [a]	Time[b]	Marker	Results	Ref.
Methylnitronitrosoguanidine	0.44	16	Auxotrophy for glycine	+	61
Ethylmethanesulfonate	110	16	Auxotrophy for glycine	+ +	61
ICR-191	0.25	16	Auxotrophy for glycine	+/−	61
Hydroxylamine		16	Auxotrophy for glycine	−	61
Caffeine	3000	16	Auxotrophy for glycine	−	61
Ultraviolet light	69 erg/mm^2		Auxotrophy for glycine	+	61
X-ray	200 rd		Auxotrophy for glycine	+/−	61
Dimethylnitrosamine	1500	16	Auxotrophy for glycine	+	62
Methylnitrosourea	75	16	Auxotrophy for glycine	+	62
Methylnitrosourethane	1.0	16	Auxotrophy for glycine	+	62
Ultraviolet light	400 erg/mm^2		6TG^R	+	46
Ethylmethanesulfonate	400	16	6TG^R	+ +	47
Methylnitrosourea	100	2.5	Cyclohexa-mide^R	+	104
Methylnitronitrosoquanidine	0.2	16	6TG^R	+ +	77,78
ICR-191	3	16	6TG^R	+	77,78
X-ray	800 rd		6TG^R	+	77,78
Dimethylnitrosamine	1300	2	6TG^R	+	77,78

Hamster Lung (V-79) Cells

Agent	Dose	Time	Marker	Results	Ref.
Triflourothymidine	0—0.4	16	8-Azaguanine resistance	−	50
Fluorodeoxyuridine	0—0.4	16	8-Azaguanine resistance	−	50
Hydroxyurea	0—4	16	8-Azaguanine resistance	−	50
Bromodeoxyuridine (BUdR)	0—0.05	16	8-Azaguanine resistance	+ +	50
Arbinofuranocytosine	0—0.0004	16	8-Azaguanine resistance	+ +	50
X-Ray	0—1000 rd		8-Azaguanine resistance	+ + +	15
Ultraviolet light	0—210 erg/mm^2		8-Azaguanine resistance	+ +	15
Acetylaminofluorene	0—20	4	8-Azaguanine resistance	−	49
N-Hydroxyacetylaminofluorene	0—20	4	8-Azaguanine resistance	+	49
N-Acetoxyacetylamino-fluorene	0—4	4	8-Azaguanine resistance	+ + +	49
Ethylmethanesulfonate	10 mM	2	8-Azaguanine resistance	+ +	21
Methylmethanesulfonate	1 mM	2	8-Azaguanine resistance	+ +	21

TABLE 1 (continued)

In Vitro Mammalian Mutagenesis

Hamster Lung (V-79) Cells

Agent	Dose[a]	Time[b]	Marker	Results	Ref.
Methylnitronitrosoguanidine	0.01 mM	2	8-Azaguanine resistance	+ +	21
Methylnitronitrosoguanidine	0—0.0008 mM	3	8-Azaguanine resistance	+ +	48
Benzanthracene (BA)	0—0.03 mM	3	8-Azaguanine resistance	−	48
BA-k[c]-epoxide	0—0.025 mM	3	8-Azaguanine resistance	+ +	48
BA-k-phenol	0—0.017 mM	3	8-Azaguanine resistance	+	48
BA-cis-dihydrodiol	0—0.030 mM	3	8-Azaguanine resistance	−	48
BA-trans-dihydrodiol	0—0.045 mM	3	8 Azaguanine resistance	−	48
3-Methylcholanthrene (MCA)	0—0.035 mM	2	8-Azaguanine resistance	+	48
MCA-k-epoxide	0—0.025 mM	2	8-Azaguanine resistance	+ + +	48
MCA-k-phenol	0—0.035 mM	2	8-Azaguanine resistance	+	48
MCA-cis-dihydrodiol	0—0.035 mM	2	8-Azaguanine resistance	+	48
MCA-trans-dihydrodiol	0—0.035 mM	2	8-Azaguanine resistance	+	48
Dibenzanthracene (DBA)	0—0.052 mM	2	8-Azaguanine resistance	+	48
DBA-k-epoxide	0—0.050 mM	2	8-Azaguanine resistance	+ +	48
DBA-k-phenol	0—0.025 mM	2	8-Azaguanine resistance	+ + +	48
DBA-cis-dihydrodiol	0—0.048 mM	2	8-Azaguanine resistance	+	48
DBA-trans-dihydrodiol	0—0.064 mM	2	8-Azaguanine resistance	+	48
7,12-Dimethylbenzanthracene (DMBA)	0—0.019 mM	2	8-Azaguanine resistance	+	48
DMBA-k-epoxide	0—0.003 mM	2	8-Azaguanine resistance	+ + +	48
7-Bromo-DMBA	0—0.018 mM	2	8-Azaguanine resistance	+ +	48
7-Methylbenzanthracene	0—0.022 mM	2	8-Azaguanine resistance	+	48
7-Bromo, 12-methylbenzanthracene	0—0.003 mM	2	8-Azaguanine resistance	+ + +	48
Gamma radiation	1200 rd		8-Azaguanine[R] (8AG[R])	+ + +	8
X-rays	800 rd		6-Thioguanine[R] (6TG[R])	+	93
Ultraviolet light	75 erg/mm^2		6TG[R]	+	93
Ethylmethanesulfonate	$5 \times 10^{-2} M$	2	6TG[R]	+	93
Bromodeoxyuridine	$10^{-4} M$	30 min	6TG[R]	+	1
Pyrene	1	2 days	8AG[R]	−	52
	1	2 days	Oua[R]	−	52

TABLE 1 (continued)

In Vitro Mammalian Mutagenesis

Hamster Lung (V-79) Cells

Agent	Dose [a]	Time [b]	Marker	Results	Ref.
Phenanthrene	1	2 days	8AG^R	−	52
	1	2 days	Oua^R	−	52
Chrysene	1	2 days	8AG^R	−	52
	1	2 days	Oua^R	−	52
Benzanthracene	1	2 days	8AG^R	−	52
	1	2 days	Oua^R	−	52
Dibenz[a,c]anthracene	1	2days	8AG^R	+/−	52
	1	2 days	Oua^R	−	52
Dibenz[a,h]anthracene	1	2 days	8AG^R	+/−	52
	1	2 days	Oua^R	−	52
7-Methylbenzanthracene	1	2 days	8AG^R	+/−	52
	1	2 days	Oua^R	+	52
3-Methylcholanthrene	1	2 days	8AG^R	+ +	52
	1	2 days	Oua^R	+	52
	1	2 days	TS reversion	+	52
Benzo(α)pyrene	1	2 days	8AG^R	+ +	52
	1	2 days	Oua^R	+	52
	1	2 days	TS Reversion	+	52
7,12-Dimethylbenz(α)-anthracene	0.1	2 days	8AG^R	+ +	52
	0.1	2 days	Oua^R	+/−	52
	0.1	2 days	TS Reversion	+	52
Dimethylnitrosamine	50mM	1	8AG^R	+ +	66
Diethylnitrosamine	50mM	1	8AG^R	+ +	66
Dipropylnitrosamine	20mM	1	8AG^R	+	66
Dibutylnitrosamine	1.0mM	1	8AG^R	+	66
Dipentylnitrosamine	1.0mM	1	8AG^R	+	66
Methyl-t-butylnitrosamine	50mM	1	8AG^R	−	66
Diphenylnitrosamine	0.5mM	1	8AG^R	−	66
Methylphenylnitrosamine	5.0mM	1	8AG^R	−	66
Nitrosomorpholine	50mM	1	8AG^R	+	66
Nitrosopyrrolidine	50mM	1	8AG^R	+	66
Nitrosomethylpiperazine	50mM	1	8AG^R	+	66
Aflatoxin B$_1$	0.8μM	3	6TG^R	+	65
Benzo(α)pyrene	25μM	3	6TG^R	+	65
3-Methylcholanthrene	100μM	3	6TG^R	+	65
7,12-Dimethylbenzanthracene	70μM	3	6TG^R	+	65
Dibenz[a,h]anthracene	80μM	3	6TG^R	+/−	65
Dibenz[a,c]anthracene	80μM	3	6TG^R	+	65

L5178Y Mouse Lymphoma Cells

Agent	Dose [a]	Time [b]	Marker	Results	Ref.
Methylnitronitrosoguanidine	$5 \times 10^{-7} M$	2	ASN$^-$→ASN$^+$	+/−	90
Ethylmethanesulfonate	$1 \times 10^{-2} M$	2	ASN$^-$→ASN$^+$	+/−	90
ICR-372	$2 \times 10^{-6} M$	2	ASN$^-$→ASN$^+$	+/−	90
ICR-191	$5 \times 10^{-6} M$	2	ASN$^-$→ASN$^+$	+/−	90
Hycanthonemethanesulfonate	90μM	2	BUdR^R	+	23
Lucanthone HCl	40μM	2	BUdR^R	+	23
IA3	70μM	2	BUdR^R	+/−	23
IA4	40μM	2	BUdR^R	+	23
IA5	20μM	2	BUdR^R	+ +	23
IA6	80μM	2	BUdR^R	+ +	23
X-rays	600rd		6TG^R	+ +	63
Ethylmethanesulfonate	$5 \times 10^{-3} M$	2	6TG^R	+ +	63

TABLE 1 (continued)

In Vitro Mammalian Mutagenesis

P388 Mouse Lymphoma Cells

Agent	Dose	Time	Marker	Results	Ref.
Nitrogen mustard	$0.32\mu M$	30 min	Thymidine^R	+	7
Methylenedimethanesulfonate	$0.25mM$	30 min	Thymidine^R	+	7
Methylnitrosourethane	$1.0mM$	30 min	Thymidine^R	+	7
Methylmethanesulfonate	$2.18mM$	30 min	Thymidine^R	+	7
X-rays	?		Thymidine^R	+	7
Ultraviolet light	?		Thymidine^R	+	7

S49 Mouse Lymphoma Cells

Agent	Dose	Time	Marker	Results	Ref.
Ethylmethanesulfonate	800	24	Bt$_2$cAMP^R	+/−	36
	800	24	6TG^R	+/−	36
Methylnitronitrosoguanidine	3	24	Bt$_2$cAMP^R	+	36
	3	24	6TG^R	+	36
ICR-191	1.2	24	Bt$_2$cAMP^R	−	36
	1.2	24	6TG^R	+	36

Human Fibroblast Strain HF10

Agent	Dose	Time	Marker	Results	Ref.
X-rays	350 rd		fruc$^-$→fruc$^+$	+/−	25

Human Fibroblasts (Secondary Culture)

Agent	Dose	Time	Marker	Results	Ref.
N-Acetoxy-AAF	$6\mu M$	4	8AG^R	+	67

Human Fetal Lung Fibroblasts

Agent	Dose	Time	Marker	Results	Ref.
Ethylmethanesulfonate	150	24	Oua^R	+/−	17
Methylnitronitrosoguanidine	1.5	4	Oua^R	+/−	17
Ultraviolet light	Unknown irr.	15 sec	Oua^R	+/−	7

Human Lymphoblasts; Line PGLC-33H

Agent	Dose	Time	Marker	Results	Ref.
Methylnitrosourea	12	24	6TG^R	+ +	92
Methylnitronitrosoguanidine	45 ng/ml	24	6TG^R	+	79
ICR-191	$1.25\mu M$	24	6TG^R	+	27
Chlorodeoxyuridine	$200\mu M$	24	6TG^R	+	80
Bromodeoxyuridine	$200\mu M$	24	6TG^R	+	80
Iododeoxyuridine	$75\mu M$	24	6TG^R	+	80
Heat	45°CμM	15 min	6TG^R	+	43
9-Aminoacridine	3	24	6TG^R	−	28
Ultraviolet light	120 erg/mm^2		6TG^R	+ +	101
Ethylmethanesulfonate	$0—200\mu M$	24	6TG^R	+ + +	101
Methylnitronitrosoguanidine	$0—0.1\mu M$	2	6TG^R	+ + +	101

Note: Using most mutagenic level tested: + > 10^{-4}, < 10^{-3} mutants per survivor; + + > 10^{-3}, < 10^{-2} mutants per survivor; +/− > 10^{-5}, < 10^{-4} mutants per survivor; − < 10^{-5} mutants per survivor; and + + + > 10^{-2} mutants per survivor.

- [a] In micrograms per milliliter unless otherwise specified.
- [b] In hours unless otherwise specified.
- [c] k region of DBA is the 5, 6; of 3 MC, is the 11, 12.

allow fixation of the mutagenic lesion and expression of mutant phenotype. Finally, selective conditions are applied, and mutant colonies are scored after growth to macroscopic size. In order to determine the toxicity of a treatment, cells are also cultivated under nonselective conditions immediately after treatment.

For work with human cells, some technical advantages can be exploited by the use of cells that grow freely in suspension cultures and form clones in soft agar, as opposed to anchorage-dependent fibroblasts. The principal reason for this advantage is that treatment with mutagenic agents often results in significant toxicity. Since (1) induced mutant fractions are usually less than 10^{-3}, (2) a limit of 10^4 fibroblasts per petri dish has been reported, and (3) surviving fractions range as low as 0.10, one will observe only $10^{-3} \times 10^4 \times 0.10 = 1$ mutant clone per petri dish, assuming 100% plating efficiency. In practice, this means that several hundred petri dishes must be handled for each experimental point.[2,67] Equal statistical significance will be provided using four petri dishes when human lymphoblasts are used under identical treatment conditions.

A second important technical problem arises from the cellular biochemistry of 6TG and 8AG resistance. While early workers recognized that an "expression period" of growth under nonselective conditions would be required, first, to allow for "fixation" of the genetic lesion and, second, to allow the mutated cell to rid itself (presumably via degradation and/or dilution) of preexisting gene products, the almost exclusive use of anchorage-dependent cells greatly retarded determination of the true "optimal" time for applying selective conditions. The literature is rife with time-of-expression curves, which rise to a maximum (usually within 48 to 96 hr) and then, as plates approach confluency, begin to decline.[15,21,82] This decline, while obviously not a characteristic generally associated with a true mutational event, can be explained in terms of "metabolic cooperation", i.e., in the case of 6TG resistance, the transfer of 6-thiodeoxyguanylic acid from nonmutant to mutant cells in close contact with one another.[24,37] Finally, Chasin[20] reported the appearance of a stable mutant fraction in CHO cells requiring 7 days of continuous growth under nonselective conditions following treatment with mutagen. In the past year, three laboratories have confirmed these results and extended them to other rodent cell lines.[1,46,93] Another group[63] has reported an extended expression time for 6TG resistance in L5178Y cells, but has also reported an unexplained decline in mutant fraction after 11 days.

The unexpected discovery of an extended period of phenotypic lag before all mutants induced by treatment express the mutant phenotype has opened the quantitative observations of early students of mammalian cell mutagenesis (including ourselves)[91] to serious question.

We have replotted data from the report of Chasin,[20] who first documented this long phenotypic lag for resistance to 6-thioguanine in Chinese hamster ovary (CHO) cells[20,77,78] and the report of Cox and Masson,[25] who reported the analogous result for diploid human fibroblasts, as well as the report of Van Zeeland and Simons[93] for the V79 hamster cell fibroblast line and our own demonstration[92] of this remarkably long phenotypic lag in diploid human lymphoblasts (Figures 1 to 4).

Despite the importance of characterizing the phenotypic lag, this parameter is often not examined or the results are often not reported. Representative reports of concentration or dose response from several laboratories are shown here to demonstrate the kind of data to be expected from mammalian gene locus mutation assays (Figures 5 to 7).

Caveat — Cell culture for even the simplest assays requires an experienced staff, alert to the many details necessary to success on this interface between art and science. A centrally funded industry effort is recommended to competent laboratories willing to undertake studies simplifying and even automating these kinds of assays. Our laboratory at MIT is given to the study of cells only of human origin, because of the belief that the mutational response of cells in vitro to a wide range of chemicals will reflect the species of origin. At this time, there are insufficient data to confirm or reject this hypothesis, except in the related area of DNA repair.

B. Bacterial Gene Locus Mutation Assay

The earliest reported bacterial test system used streptomycin-dependent strains of *Esche-*

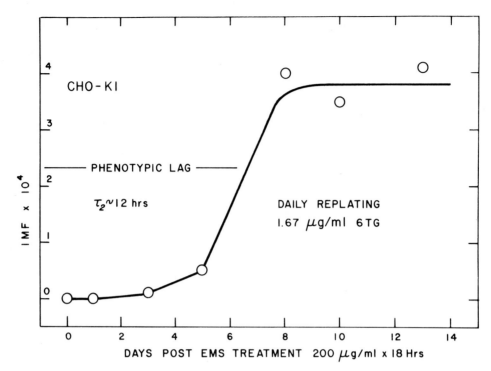

FIGURE 1. Phenotypic lag in mammalian cells. By plotting induced mutant fraction vs. time after treatment until plating, one can see a phenotypic expression time of about ten generations for CHO-K1. (From Chasin, L. A., *J. Cell. Physiol.*, 82, 229, 1973. With permission.)

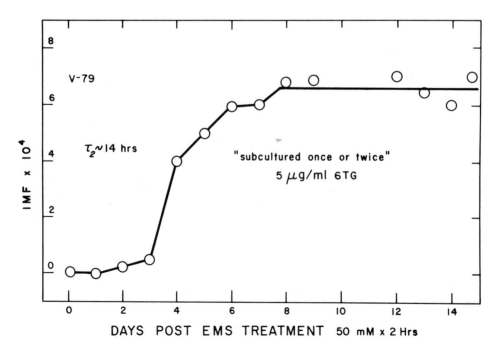

FIGURE 2. Phenotypic lag in mammalian cells. By plotting induced mutant fraction vs. time after treatment until plating, one can see a phenotypic expression time of about ten generations for V-79. (From Van Zeeland, A. A. and Simons, J. W. I. M., *Mutat. Res.*, 35, 129, 1976. With permission.)

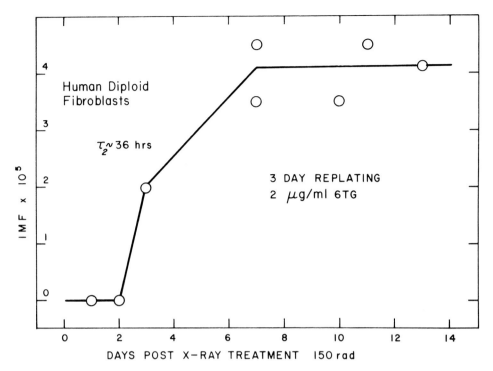

FIGURE 3. Phenotypic lag in mammalian cells. By plotting induced mutant fraction vs. time after treatment until plating, one can see a phenotypic expression time of about five generations for human fibroblasts. (From Cox, R. and Masson, W. K., *Mutat. Res.*, 37, 125, 1976. With permission.)

richia coli which were mutated to streptomycin-independence.[14] Using this sytem, Demerec et al.[29] tested numerous chemical compounds, reporting positive results with hydrogen peroxide, formaldehyde, phenol, acriflavine, and caffeine and negative results with sulfuric acid, hydrochloric acid, and sodium salicylate (selected results).

Mutation assays generally have two forms. The streptomycin-independence test is an example of a *reverse mutation assay* in which a mutant strain is reverted to its parent phenotype. The other common type of assay is the forward mutation assay, in which a wild-type strain is mutated to some measurable condition, usually to antibiotic or other drug resistance.

There are relatively few instances in the literature where forward mutation has been used to screen potential mutagens. Elespuru et al.[31] measured increased mutation to novobiocin resistance in *Haemophilus influenzae* after treatment with methylnitronitroso guanidine, methylnitrosourethane, and nitrosocarbaryl. Others have used resistance to colicin B in *E. coli*.[54,55] Many bacteria are sensitive to 5-methyltryptophan, and a mutation in any one of five genes can result in a resistant strain.[32,68] In all of these cases, the assay consists simply of plating treated cells in the presence of the drug and scoring the resulting resistant colonies.

Mutation of streptomycin dependence (S_D) to streptomycin independence (S_I) as an assay system was improved by Inyer and Szybalski.[53] Their method involved spotting a paper disc containing the compound to be tested on a lawn of S_D *E. coli* plated in minimal medium. β-Propiolactone, azaserine, and triethylenemelamine (among others) increased the number of revertant S_I colonies.[53] Later investigators discovered that while base-pair substitution mutagens were scored in this system, frameshift mutagens (ICR-191) did not cause reversion.[84]

A series of *E. coli* mutant strains auxotrophic for tryptophan have been developed, which revert to prototrophy when treated with a variety of mutagens.[13,45,105] Likewise, there are specific arginine-requiring *E. coli* strains that can be reverted to prototrophy.[31,55,64] These strains often carry a particular DNA repair deficiency, which makes them more sensitive to some mutagens (indicating that mutational damage can be re-

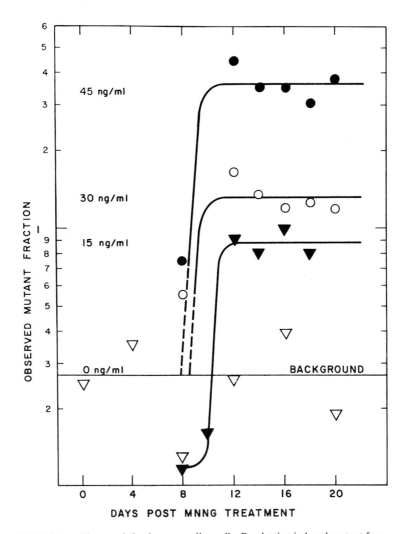

FIGURE 4. Phenotypic lag in mammalian cells. By plotting induced mutant fraction vs. time after treatment until plating, one can see a phenotypic expression time of about 12 generations for human lymphoblasts. (From Thilly, W. G., *Toxicol. Environ. Health*, 2, 1343, 1977. With permission from Hemisphere Publishing Corporation.)

paired), but less sensitive to others (suggesting that faulty repair can increase mutation frequency).[55] There are numerous other mutant strains that can be reverted to prototrophy, such as *E. coli* lac⁻ to lac⁺ [100] or met⁻ to met⁺,[87] but most of these are only able to detect a narrow range of mutagens (the lac and met respond only to acridines among the compounds tested).[87,100]

The most widely used bacterial mutation screening-system is the *Salmonella typhimurium* test, also known as the Ames System.[6] Commonly, four tester strains of histidine-requiring *S. typhimurium* mutants are tested for reversion to prototrophy with a potential mutagen. Each of these strains carries a different mutation in the histidine operon; each lesion has a different molecular characteristic, so that a potential mutagen has a wider target spectrum to affect.

The *Salmonella* tester strains carry additional mutations: a DNA repair deletion[3] and a mutation in the lipopolysaccharide coat, which allows potential mutagens better access to the membrane[4] to increase the sensitivity to mutagens. A plasmid has been incorporated into two strains; this further enhances their sensitivity.[69]

Many chemical compounds are known to be

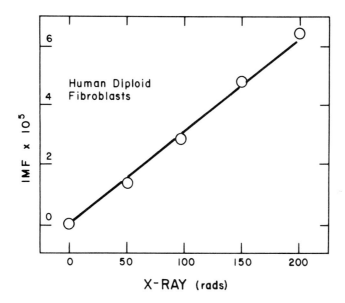

FIGURE 5. A linear dose-response curve for mutation to 6TG^R induced by X-irradiation in human diploid fibroblasts. (From Cox, R. and Masson, W. K., *Mutat. Res.,* 37, 125, 1976. With permission.)

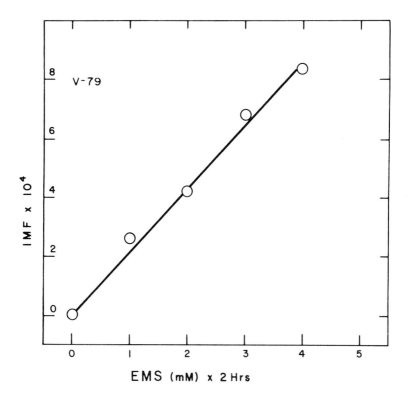

FIGURE 6. A linear dose-response curve for mutation to 6TG^R induced by ethylmethanesulfonate in V-79 hamster cells. (From Van Zeeland, A. A. and Simons, J. W. I. M., *Mutat. Res.,*35, 129, 1976. With permission.)

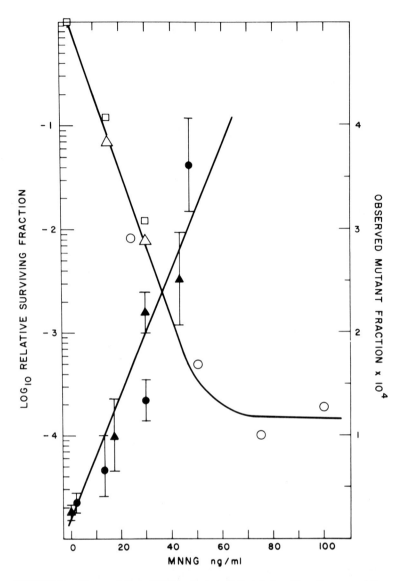

FIGURE 7. Mutagenicity to 6TG^R and cytotoxicity is induced by nitrosoguanidine in human lymphoblasts. Treatment was for 24 hr. (From Penman, B. W. and Thilly, W. G., *Som. Cell Genet.*, 2, 325, 1976. With permission from Plenum Press.)

biologically active only after biotransformation by a variety of drug-metabolizing enzymes. Therefore, a metabolizing system composed of rat liver postmitochondrial supernatant, which contains the microsomal oxidative enzymes, has been coupled to the test systems.[5]

The earliest results obtained with the *Salmonella* system were gathered with the use of the "spot test". This procedure involved "spotting" a solid crystal or drop of a compound on a lawn of mutant bacteria.[3] As the compound diffused radially outward from its point of application, a "zone" of inhibition, presumably caused by toxic effects, was observed. This was surrounded by a ring of revertant colonies.[3]

Later, the procedure was changed so that dosing was more uniform. Some experiments were performed by adding a known concentration of mutagen onto a preplated lawn of bacteria. Others were performed by mixing a known concentration of a chemical with the bacteria in suspension and immediately plating the dosed suspension onto agar plates. The latter procedure was the one followed when using the metabolizing system; in that case, bacteria, chemical, and postmitochondrial supernatant were mixed together in suspension and then plated onto agar plates.[6]

In the Ames procedures described above, bacteria were plated in the presence of a trace of histidine. This allowed each bacterium to divide a certain undefined number of times before histidine became limiting.[6] This was done to permit mutation expression time, which is estimated to be one or two generations.

Many chemicals have been tested for mutagenicity with this system. McCann et al.[70] have obtained positive results with aromatic amines, alkyl halides, polycyclic aromatics, esters, epoxides, carbamates, nitrogen-containing aromatics and heterocyclics, nitrosamines, fungal toxins, antibiotics, and azo and diazo compounds. Numerous other laboratories have since published much data obtained with the test. Diverse classes of compounds have been examined, including nitrosamines,[12,98] nitrofurans,[97] quinolines,[33,73] anthroquinines,[16] polycyclic hydrocarbons,[44] azodyes,[38] vinyl chloride,[39] fungal metabolites,[96] pesticides,[83] food additives,[94] drugs,[88] and even the outer surface of charred beefsteak.[74]

The *E. coli* system described by Ellenberger and Mohn[32] seems well suited to the general screening of chemicals for mutagenesis. *E. coli* K-12/galR$^s_{18}$/arg$_{56}$/nad$_{113}$ is a biauxotrophic mutant that requires arginine and nicotinic acid. Since it carries a gal⁻ phenotype and is also sensitive to 5-methyltryptophan inhibition, this single strain can be tested in four ways for mutation:

1. Reversion of galR$^s_{18}$ mutation to gal⁺ — known to be the result of a second forward mutation
2. Forward mutation to 5-methyltryptophan resistance
3. Back mutation to arg⁺
4. Back mutation to nad⁺

This system has not been used for any extensive screening of environmental mutagens.

Our laboratory at MIT has recently adapted a guanine-analogue resistance mutation assay so that it may be used with *S. typhimurium*.[85] Using a strain developed by Ames et al.[4] with the deep rough mutation permits many chemicals easier access through the cell membrane. Thus, one is now able to perform quantitative forward mutation assays for the loss of the enzyme xanthine-guanine phosphoribosyltransferase (XGPRT) with remarkable facility. (At least it seems remarkable to us mammalian cell culture workers.)

An outline of the procedure, based upon the reasoning that cell death is a response independent of cell mutation, is given in Figure 8. A simple comparison of the genetic endpoints between Ames' reversion assay and our suggested forward assay is shown in Figure 9.

The background mutant fraction for any mutation assay fluctuates widely among days, as can be seen in the following seven summaries for five reversion and two forward mutation assays in *S. typhimurium* (Figure 10). As a result, for example, in TA100, background levels of 6 or 42×10^{-8} revertants per surviving cell are both within the 95% confidence limits. One might erroneously conclude that a background of 10×10^{-8} and a treated culture result of 30×10^{-8} represented a three times increase in mutant fraction, when no mutation occurred at all! Roughly speaking, an induced mutant fraction should rise to greater than 2 SD above the mean background before it is considered statistically significant with 95% confidence. In the worst possible case represented by TA1538, this would mean a result on a treated culture of greater than $2.4 \text{ (mean)} + 2[2.5 \text{ (SD)}] = 7.4 \times 10^{-8}$ would be necessary for the rejection of the null hypothesis (that no mutation occurred). Prudence dictates that a somewhat greater confidence limit should be used, however. We have achieved this in the following presentation of data, by noting that for all assays, when observations on treated plates are corrected for background observed in that experiment, if the resulting value is greater than twice the mean background for the assay, the result may always be accepted as statistically significant with greater than 95% confidence. This tactic also permits facile comparison of assays for sensitivity, since the parameter IMF/B ([observed mutant fraction — observed background fraction]/ mean background for assay) is automatically normalized and has essentially the same value, 2.0, for significance in each assay.

The following studies have been completed, and are published.[86] The studies together show that a single *S. typhimurium* strain TM677, an

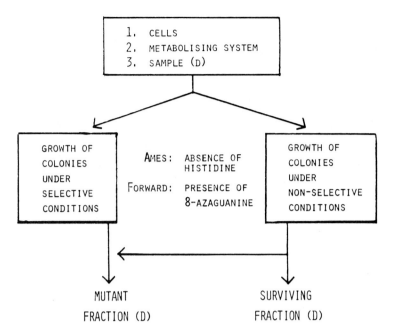

FIGURE 8. Quantitation of mutation requires a clean selection system and an independent means of measuring cell survival. The forward and reverse assays are performed in the same way when done quantitatively. The only difference is in the selective conditions.

his⁺ revertant of TA1535 carrying the R-factor plasmid, can be used to replace the set of five reversion assays, since it detects all the compounds of varying chemical and mutagenic classes with sensitivity essentially equal to the most sensitive of the reversion assays for a particular compound (Figures 11 to 15 and Table 2). This approach is simpler and easier by virtue of the use of a single strain. Also, very small test volumes may be used, which is important when dealing with compounds of limited quantity.

C. Drug Metabolism in Mutation Assays

The variety of enzyme systems that react with potential mutagens and the differences among the many cell types in the human body severely complicate the job of predicting the genetic hazard of a particular environmental pollutant.[106] Before pronouncing a substance either safe or dangerous to man, one must not only consider the characteristics of the chemical in its natural form, but also those of all its metabolic by-products.

Rather than defining the in vivo metabolites of a particular chemical of interest and then testing each of them separately, it is far easier to simply coincubate drug-metabolizing elements, the target cells (which usually do not have any drug-metabolizing capacity), and the parent compound. In mammalian cells, Krahn and Heidelberger[65] were the first to use the postmitochondrial supernatant from rats; they obtained increased mutant fractions with aflatoxin B_1 (Figure 16) and a series of polycyclic hydrocarbons. Earlier, Huberman and Sachs[51,52] had activated polycyclic hydrocarbons by cocultivating V79 with hamster embryo cultures; however, these studies utilized selection to 8AG resistance after only 48 hr, which, as discussed above, may have been insufficient. Some investigators have also used a host-mediated assay. For example, O'Neill et al.[78] injected CHO hamster cells into nude mice and obtained an increase in the $6TG^R$ mutant fraction after an intraperitoneal injection of dimethylnitrosamine.

In bacterial systems, a postmitochondrial supernatant is generally used as the activating element;[6,85] host-mediated systems are also used.

D. Gene Structure Affects the Probability of Observing Mutation

From the perspective of researchers inter-

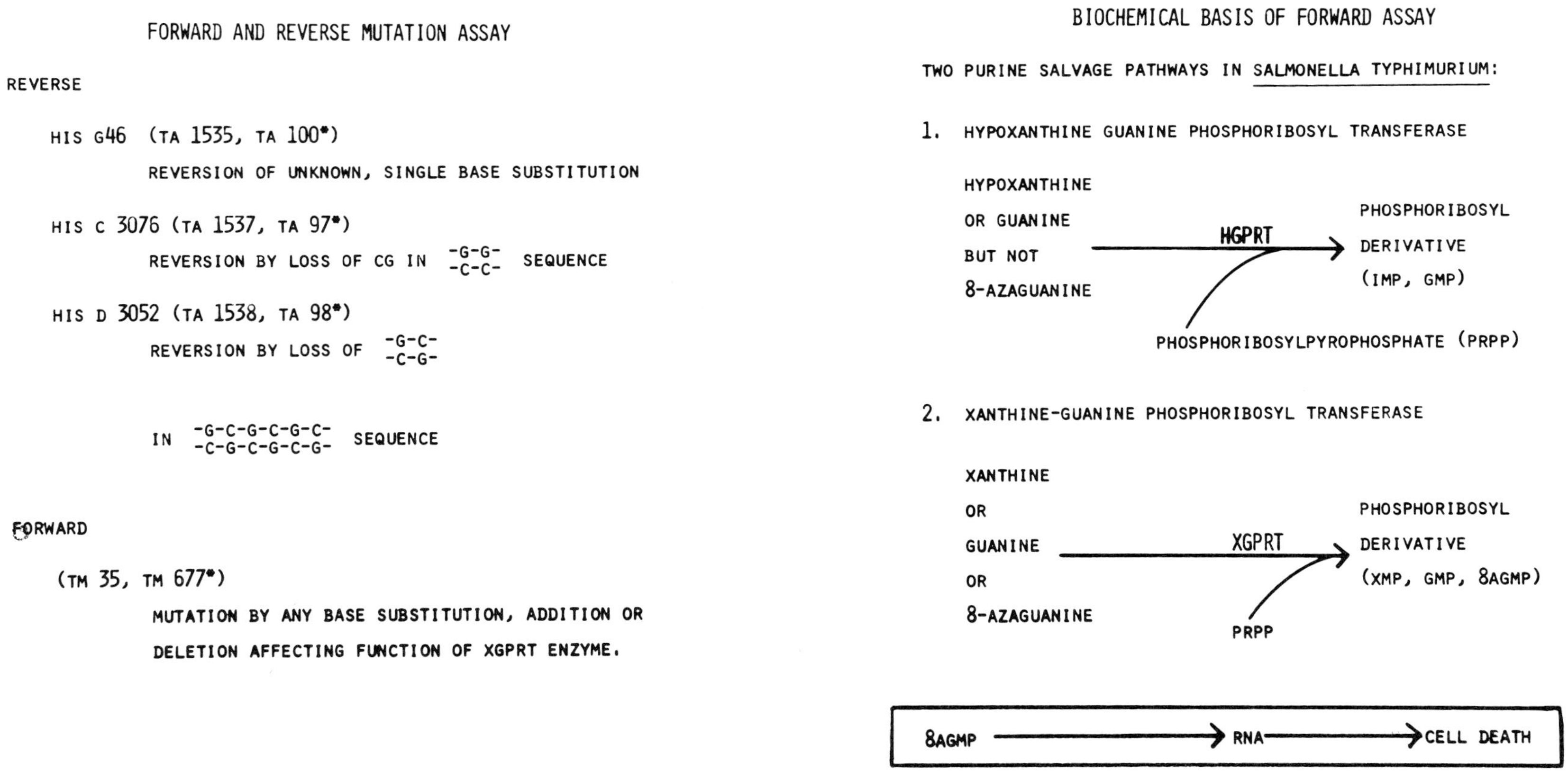

FORWARD AND REVERSE MUTATION ASSAY

REVERSE

HIS G46 (TA 1535, TA 100*)

REVERSION OF UNKNOWN, SINGLE BASE SUBSTITUTION

HIS C 3076 (TA 1537, TA 97*)

REVERSION BY LOSS OF CG IN $\begin{array}{c}-G-G-\\-C-C-\end{array}$ SEQUENCE

HIS D 3052 (TA 1538, TA 98*)

REVERSION BY LOSS OF $\begin{array}{c}-G-C-\\-C-G-\end{array}$

IN $\begin{array}{c}-G-C-G-C-G-C-\\-C-G-C-G-C-G-\end{array}$ SEQUENCE

FORWARD

(TM 35, TM 677*)

MUTATION BY ANY BASE SUBSTITUTION, ADDITION OR DELETION AFFECTING FUNCTION OF XGPRT ENZYME.

BIOCHEMICAL BASIS OF FORWARD ASSAY

TWO PURINE SALVAGE PATHWAYS IN *SALMONELLA TYPHIMURIUM*:

1. HYPOXANTHINE GUANINE PHOSPHORIBOSYL TRANSFERASE

HYPOXANTHINE
OR GUANINE →[HGPRT] PHOSPHORIBOSYL DERIVATIVE (IMP, GMP)
BUT NOT
8-AZAGUANINE

PHOSPHORIBOSYLPYROPHOSPHATE (PRPP)

2. XANTHINE-GUANINE PHOSPHORIBOSYL TRANSFERASE

XANTHINE
OR
GUANINE →[XGPRT] PHOSPHORIBOSYL DERIVATIVE (XMP, GMP, 8AGMP)
OR
8-AZAGUANINE

PRPP

8AGMP → RNA → CELL DEATH

FIGURE 9. Forward and reverse mutation assays.

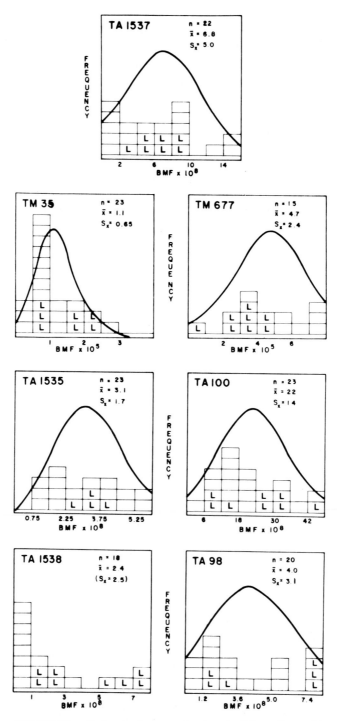

FIGURE 10. Background mutant fractions in *Salmonella typhimurium*. Each block represents a separate determination of the background mutant fraction on different days.

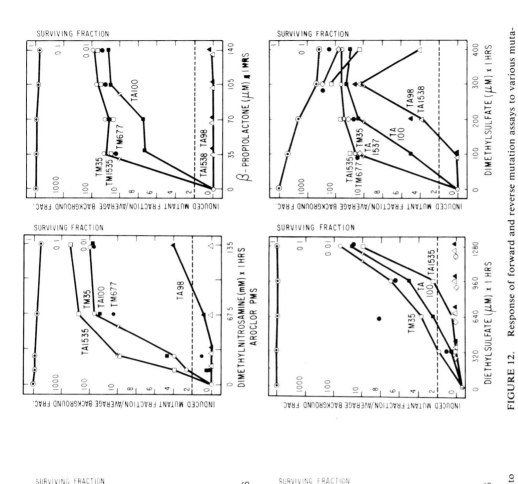

FIGURE 12. Response of forward and reverse mutation assays to various mutagens.

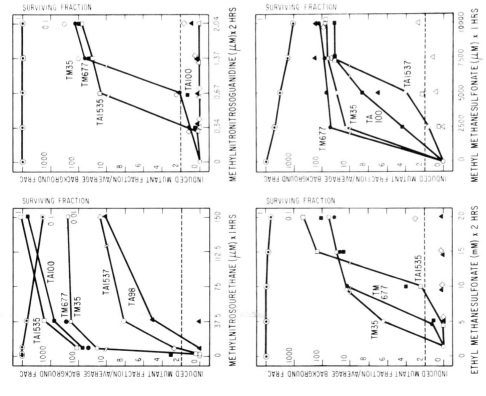

FIGURE 11. Response of forward and reverse mutation assays to various mutagens.

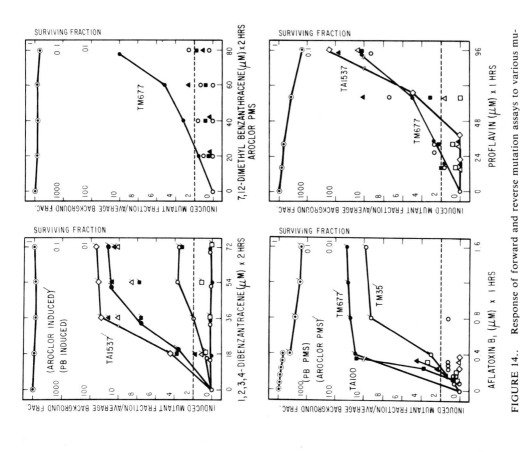

FIGURE 14. Response of forward and reverse mutation assays to various mutagens.

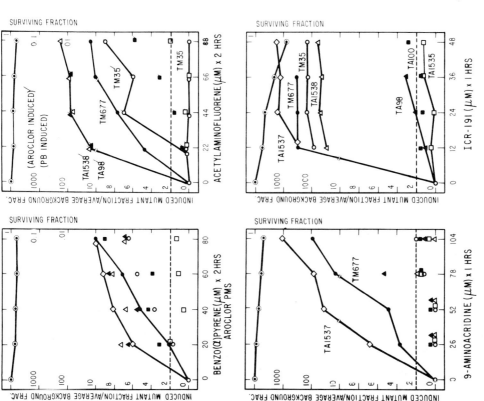

FIGURE 13. Response of forward and reverse mutation assays to various mutagens.

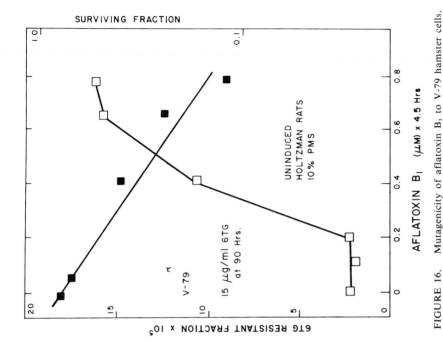

FIGURE 16. Mutagenicity of aflatoxin B$_1$ to V-79 hamster cells. Incubation of aflatoxin, postmitochondrial supernatant (PMS), and V-79 resulted in induction of mutations to 6TG^R. Without the PMS, there was no increase. (From Krahn, D. and Heidelberger, C., *Mutat. Res.*, 46, 27, 1977. With permission.)

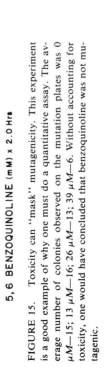

FIGURE 15. Toxicity can "mask" mutagenicity. This experiment is a good example of why one must do a quantitative assay. The average number of colonies observed on the mutation plates was 0 μM—15; 13 μM—16; 26 μM—13; 39 μM—6. Without accounting for toxicity, one would have concluded that benzoquinoline was not mutagenic.

TABLE 2

Forward and Reverse Assays in *Salmonella typhimurium*

Compound	Revertant Strain[a]	R[b]	TM35[b]	TM677[b]
Methylmethanesulfonate (mM)	TA100	1	0.5	0.5
Ethylmethanesulfonate (mM)	TA100	7	2.5	6
Dimethylsulfonate (μM)	TA1535, TA1537	20	20	20
Diethylsulfonate (μM)	TA100	600	320	400
N-Methyl-N-nitro-nitrosoguanine (μM)	TA1535	350	670	670
Methylnitrosourethane (μM)	TA1535, TA100	4	4	4
Dimethylnitrosamine (μM)	TA1535	10	15	30
β-Propiolactone (μM)	TA1535	100	100	100
9-Aminoacridine (μM)	TA1537	10	100	15
Proflavine (μM)	TA98	30	30	30
ICR-191 (μM)	TA1537, TA1538	4	4	4
Aflatoxin B$_1$ (μM)	TA100	0.2	0.3	0.2
1,2,3,4-Dibenzanthracene (μM)	TA1537, TA100	10	36	12
7,12-Dimethylbenzanthracene (μM)	TA98	60	80	30
Acetylaminoflourene (μM)	TA1538, TA98	5	25	8
Benzo(A)pyrene (μM)	TA1538, TA98, TA1537	10	25	20

Note: Sensitivity is here defined as the lowest concentration to which an assay responds positively at the 99% confidence limit. Table 2 shows that the largest difference between the two assays was a factor of two; the forward and reverse assays are equisensitive.

[a] Most sensitive revertant strain.
[b] Concentration at which significant mutation is detectable.

ested in mechanisms of mutation, DNA has three levels of organization: base pairs, triplet codons, and unique recognition sequences. Each level is a molecular means of encoding a certain kind of information and insuring that the information is faithfully inherited. Disruption of the code at any of the three levels represents a genetic mutation.

The first two levels are familiar to all students of molecular biology and need not be reviewed here. The existence of unique recognition sequences is a more recent finding arising from theoretical [40] and now experimental research [41] concerning the structure of the operator region for the lac operon in *E. coli*. The classical operator region of Jacob and Monod[58] was the DNA sequence preceding the structural genes for the enzymes of the lac operon to which the repressor protein was bound. When repressor protein was bound to the operator, the lac operon was "off" and vice versa.

The information contained in such an operator must perforce be contained in the linear sequence of base pairs of that region. If an operon is to be uniquely controlled by a specific repressor protein, there is a lower limit on the amount of information which the operator sequence must contain. In *E. coli,* it has been calculated that in order for a linear sequence to be unique, some 12 to 15 base pairs are required.[40] In human cells containing larger amounts of DNA, we have calculated a lower limit of about 20 base pairs for a sequence to be unique (less than 1% chance of redundancy).

More recent experimental studies of the lac operator region[42] have found it to have a twofold axis of symmetry and a length of 24 base pairs. The corresponding repressor protein also has a twofold symmetry.[76] Binding of repressor to the operator involves only 12 of the 24 base pairs, in close agreement with theoretical predictions.

Mutations are hereditable changes in DNA and can arise by mistakes in replication, by the direct action of physical or chemical agents, or by errors in enzymatic repair of damaged DNA. Mutations in DNA are translated into mistakes in protein, which often result in loss of enzymatic or structural function. Mutations can be classified into four types: base pair substitu-

TABLE 3

The Genetic Code

UUU	phe	UCU	ser	UA*U*	tyr[a]	UGU	cys[a]
UUC	phe	UCC	ser	UA*C*	tyr[a]	UGC	cys[a]
U*U*A	leu[a]	U*C*A	ser[a]	UAA	TERM	UGA	TERM
U*U*G	leu[a]	U*C*G	ser[a]	UAG	TERM	U*G*G	tryp[a]
CUU	leu	CCU	pro	CAU	his	CGU	arg
CUC	leu	CCC	pro	CAC	his	CGC	arg
CUA	leu	CCA	pro	CAA	glun[a]	*C*GA	arg[a]
CUG	leu	CCG	pro	CAG	glun[a]	CGG	arg
AUU	ileu	ACU	thr	AAU	aspn	AGU	ser
AUC	ileu	ACC	thr	AAC	aspn	AGC	ser
AUA	ileu	ACA	thr	AAA	lys[a]	AGA	arg[a']
AUG	meth	ACG	thr	AAG	lys[a]	AGG	arg
GUU	val	GCU	ala	GAU	asp	GGU	gly
GUC	val	GCC	ala	GAC	asp	GGC	gly
GUA	val	GCA	ala	GAA	glu[a]	GGA	gly[a]
GUG	val	GCG	ala	GAG	glu[a]	GGG	gly

[a] Those codons which can be converted to termination codons by a single base-pair substitution. The genetic code is listed as it appears on mRNA. Base-pair conversions leading to termination are in italics.

tions, chain terminations, frameshifts, and large deletions.

A base-pair substitution-mutation is the replacement of one nucleotide pair with another. It is important to remember that there are six possible interconversions: two transitions (purine: pyrimidine → purine: pyrimidine; AT→ GC and GC→ AT) and four transversions (pur: pyr→ pyr: pur; GC→ CG, GC→ TA, AT→ CG, AT→ TA). Any of these substitutions results in a change in one base in an mRNA codon and can cause a change of one amino acid in the protein for which the gene codes. For example, in *E. coli,* the tryptophan synthetase A gene can be mutated by ultraviolet light. A GC→ AT transition in the DNA causes a G→ A change in the mRNA codon. The change that occurs in the protein is glycine (*G*GA)→ arginine (*A*GA). This change converts the bacterium to a tryptophan auxotroph.[99]

Base-pair substitution-mutations can be reverted in two ways. The first is by simple back mutation — a reversal of the original mutation. Yanofsky et al.[99] observed this for the above example (arginine [AGA]→ glycine [*G*GA]). The second means of reversion is a second-site mutation or suppressor, which can occur in several forms. For our example, tryptophan synthetase, a transversion in the same codon (arginine [AG*A*]→ serine [AG*Pyr*]) restores full activity. This second-site mutation is only two base pairs from the original lesion. Apparently the enzyme functions equally well with glycine or serine in this particular position, but not at all with arginine. Also, isoleucine can replace arginine at this site by another transversion (arginine [A*G*A]→ isoleucine [A*U*A]); this restores only partial activity.[99] A summary of the genetic code has been reproduced for easy reference (Table 3).

Another form of a suppressor mutation involves a mutation in a minor tRNA species. For example, if a minor tRNA of arginine (anticodon of UCU) became charged with glycine or serine instead of arginine, a certain percentage of normal active tryptophan synthetase A could be synthesized, because that tRNA would insert glycine (serine) where it reads an arginine codon.[19]

Because of the degeneracy of the genetic code, some base-pair substitution-mutations have no effect. Most transitions of the third nucleotide pair do not result in amino acid changes (the only exceptions are tryptophan [UG*G*]→TERMINATION [UG*A*]) and methionine ([AU*G*]→ isoleucine [AU*A*]). Also, some transversions in the third nucleotide pair have no effect (for instance, leucine is coded for by

CUx, where x is any of the four nucleotides). This point is important in that it reduces the number of base-pair substitution-mutations that will affect the amino acid content of protein.

A second class of mutations is the type of base-pair mutation which changes an amino acid codon to a chain termination or nonsense codon. Only three such codons exist: UAA, UAG, and UGA. We separate this class from base-pair changes involving amino acid substitutions for a very important theoretical reason. The study of the genetic code shows that of the 61 codons coding for amino acids, only 18 can mutate to one of the three chain termination codons by a single base-pair substitution. Of the six possible base-pair substitutions, five could be involved in a chain termination and one could not!

The four possible transversions (GC→CG, AT→TA, GC→TA, AT→CG) account for 14 of the 18 possible chain termination mutations, e.g., UAC(tyr)→UAA (TERM). One of the two transition mutations (GC→AT) is involved in the four remaining chain terminating mutations: CAA (glun)→UAA, CAG (glun)→UAG, UGG (tryp)→UGA or UAG, CGA (arg)→UGA.

The other transition mutation (AT→GC) cannot, however, under any circumstances give rise to a chain-termination mutation via a single base substitution. This is simply a result of the absence of C in the three termination codons and the fact that the only triplet codons which can become chain-termination codons via an AT→GC transition are already termination codons! Thus, mutagens that cause only AT→GC transitions will not cause chain-termination mutations. (Note: in the case of proteins that contain neither glutamine nor tryptophan and use the arginine codons CGU, CGC, CGG, AGA, AGG, but not CGA, the transition GC→AT cannot result in chain terminations. For such proteins, therefore, transitions will not result in chain termination.)

A third class of mutation is frameshift. Any lesion that shifts the reading frame of a gene by

Hypothetical Example of Frameshift Mutation and Reversion

. . . UCC/GCC/GAA/ACA/GGC/AGG . . .   mRNA

. . . ser – ala – glu – thr – gly – arg . . .   protein

Frameshift | Insertion of an AT pair into the DNA results in the insertion of A into mRNA and change in amino acid sequence

. . . UCC/G*AC/CGA/AAC/AGG/CAG/G . . .

. . . ser – asp – arg – aspn – arg – glun – . . .

Frameshift Reversion | Deletion of a CG pair from the DNA results in a deletion of C from mRNA and restoration of much of the original amino acid sequence

. . . UCC/G*AC/GAA/ACA/GGC/AGG . . .

. . . ser – asp – glu – thr – gly – arg . . . .

either inserting or deleting n (n ≠ 3) nucleotide pairs will cause a totally different protein to be synthesized from the point of mutation on.

The hypothetical example shows that all of the codons and the amino acids following the lesion are changed. If this sort of lesion occurred near the end of the gene, some or all activity might be retained; otherwise, the resulting protein would be nonfunctional.

The example also shows how a second-site mutation might revert the lesion, in the particular case, by restoring the reading frame with only one incorrect amino acid. Often, more than one amino acid can be incorrect without affecting activity too adversely.

A frameshift mutation can cause termination if the shift creates a nonsense codon. For example:

```
              * delete A
   . . . CCU/AUA/GGU/AAA              . . . CCU/UAG/GUA/AA . . .
                            ──────────►
   . . . — pro — ileu — gly — lys —   . . . — pro
```

However, the practical importance of this possibility is minor, since frameshifts and chain terminations would generally be expected to have equal inactivating effects on proteins synthesized from the mutated gene.

A fourth kind of mutation is a large deletion, which can involve loss of large portions of or even whole genes or groups of genes. Large deletions also generally cause loss of associated protein function.

The regions of DNA that do not code for protein are also subject to mutations. Mutant tRNA or rRNA may impair protein synthesis, but still permit cellular survival. Mutations in operator regions of DNA may result in constitutive synthesis or no transcription at all. However, all mutant strains used in mutagen screening systems have used impaired or repaired protein function as an endpoint.

Having carefully considered the levels of organization of DNA and how the various classes of single mutations may affect the translation of genetic information, the relevance of this information to the practical problem of assaying the mutagenic potential of environmental chemicals can now be discussed.

The first and possibly most important point to be applied is that a particular mutagen does not necessarily cause all kinds of mutations. A mutant strain that is auxotrophic because of a base-pair substitution-mutation will not be reverted by a mutagen that causes only frameshifts. Likewise, mutagens specific for base-pair substitution will not revert frameshifted strains.[75,95]

It also is evident that there is specificity within the types of mutagenic classes. Hydroxylamine causes only GC→AT transitions;[35] if a transversion or an AT→GC transition were necessary for the reversion of a specific lesion, hydroxylamine would test negatively in a reversion assay. The base analogues, bromodeoxyuridine and 2-aminopurine, cause transitions, but not transversions.[35] Simple alkylating agents apparently cause base-pair substitutions, but it is not known what specificity they demonstrate in causing mutations among the possibilities of four transversions and two transitions. However, they generally do not cause frameshift mutations.

It also is evident that structural specificity affects frameshift mutations. That is, a mutagen may be active in causing frameshifts in one DNA sequence but not another. A good example is the strong mutagenic effect of 9-aminoacridine to Ames' TA1537 strain compared to its lack of mutagenicity for the TA1538 strain. TA1537 and TA1538 each carry a frameshift mutation but in different genes necessary for histidine prototrophy. There is, of course, the possibility of suppressor mutations occurring, but these, too, are the result of specific lesions. Thus, in order to develop a complete screening assay, enough mutant strains would be needed to insure that all possible base-pair interconversions and frameshifts could be detected. To date, none of the reverse mutation assays contain sufficient genetic lesions to guarantee that all possible mutational mechanisms (and, thus, all possible mutagens) are detectable.

The simplest genetic solution to the problem of mutagen-mutation specificity is the use of forward mutation assays. The use of the gene for an entire functional protein increases the probability of sensitivity to a base-pair mutagen to near unity and commensurately increases the probability of containing DNA regions sensitive to a wide spectrum of frameshift mutagens.

III. CONCLUSIONS

From the arguments presented above, it has been concluded that forward mutation assays are more desirable than reverse assays for toxicological screening purposes, because: (1) it is not necessary to distinguish between classes of mutagens, e.g., base-pair substitution or frameshift; (2) the two types are equisensitive; and (3) forward assays being less specific, should respond to a wider range of mutagens than a set of reversion strains.

Obviously, bacterial cells differ from human cells and a response in one system will not necessarily predict the response in the other. However, the anticipated regulatory requirement for testing large numbers of environmental chemicals involves immediate application of state-of-the-art bioassay technology. Thus, bacterial assays are now an important aid in assessing the mutagenicity of a chemical. The authors feel, however, that bacterial assays are only a stopgap measure — eventually, *human* systems must be used to predict environmental hazards to *humans*. It has been suggested that assays utilizing human cells, coupled with a human liver-metabolizing system, might allow toxicologists to better predict the genetic effects of environmental chemicals on man.

REFERENCES

1. Abersold, P. M. and Burki, J. H., *Mutat. Res.*, 40, 63, 1976.
2. Albertini, R. J. and DeMars, R., *Mutat. Res.*, 18, 199, 1973.
3. Ames, B. N., in *Chemical Mutagens: Principles and Methods for Their Detection*, Hollaender, A., Ed., Plenum Press, New York, 1971, 267.
4. Ames, B. N., Lee, F. D., and Durston, W.E., *Proc. Natl. Acad. Sci. U.S.A.*, 70, 782, 1973.
5. Ames, B. N., Durston, W. E., Yamasaki, E., and Lee, F. D., *Proc. Natl. Acad. Sci. U.S.A.*, 70, 2281, 1973.
6. Ames, B. N., McCann, J., and Yamasaki, E., *Mutat. Res.*, 31, 347, 1975.
7. Anderson, D. and Fox, M., *Mutat. Res.*, 25, 107, 1974.
8. Arlett, C. F. and Potter, J., *Mutat. Res.*, 13, 59, 1971.
9. Arlett, C.F. and Harcourt, S. A., *Mutat. Res.*, 14, 431, 1972.
10. Arlett, C. F. and Harcourt, S. A., *Mutat. Res.*, 16, 301, 1972.
11. Bakay, B. and Nyhan, W. L., *Proc. Natl. Acad. Sci. U.S.A.*, 69, 2523, 1972.
12. Bartsch, H., Camus, A., and Malaveille, C., *Mutat. Res.*, 37, 149, 1976.
13. Berger, H., Brammar, W. J., and Yanofsky, C., *J. Bacteriol.*, 96, 1672, 1968.
14. Bertani, G., *Genetics*, 36, 598, 1951.
15. Bridges, B. A. and Huckle, J., *Mutat. Res.*, 10, 141, 1970.
16. Brown, J. P. and Brown, R. J., *Mutat. Res.*, 40, 203, 1976.
17. Buchwald, M., *Mutat. Res.*, 44, 401, 1977.
18. Capizzi, R. F., Papinmeister, B., Mullins, J., and Cheng, E., *Cancer Res.*, 34, 3073, 1974.
19. Carbon, J., Berg, P., and Yanofsky, C., *Cold Spring Harbor Symp. Quant. Biol.*, 31, 437, 1966.
20. Chasin, L., *J. Cell Physiol.*, 82, 299, 1973.
21. Chu, E. H. Y. and Malling, H. V., *Proc. Natl. Acad. Sci. U.S.A.*, 61, 1306, 1968.
22. Chu, E. H. Y., Brimer, P., Jacobson, K. B., and Merriam, E. V., *Genetics*, 62, 359, 1969.
23. Clive, D., *Mutat. Res.*, 26, 307, 1974.
24. Cox, R. P., Krauss, M. R., Balis, M. E., and Dancis, J., *Exp. Cell Res.*, 74, 251, 1972.
25. Cox, R. and Masson, W. K., *Nature (London)* 252, 308, 1974.
26. Day, R. S., *Nature (London)*, 253, 748, 1975.
27. DeLuca, J. G., Kaden, D. A., Krolewski, J. J., Skopek, T. R., and Thilly, W. G., *Mutat. Res.*, 46, 11, 1977.
28. DeLuca, J. G., Krolewski, J. J., Skopek, T. R., Kaden, D. A., and Thilly, W. G., *Mutat. Res.*, 42, 327, 1977.
29. Demerec, M., Bertani, G., and Flint, J., *Am. Nat.* 85, 119, 1951.
30. Demerec, M. and Hanson, J., *Cold Spring Harbor Symp. Quant. Biol.*, 16, 215, 1951.
31. Elespuru, R., Lijinsky, W., and Setlow, J. K., *Nature (London)*, 247, 386, 1974.
32. Ellenberger, J. and Mohn, G., *Arch. Toxicol.*, 33, 225, 1975
33. Epler, J. L., Winton, W., Ho, T., Larimer, F. W., Rao, T. K., and Hardigree, A. A., *Mutat. Res.*, 39, 285, 1977.
34. Fox, M. and Anderson, D., *Mutat. Res.*, 25, 89, 1974.

35. Freese, E., in *Chemical Mutagens: Principles and Methods for Their Detection,* Hollaender, A., Ed., Plenum Press, New York, 1971, 1.
36. Friedrich, U. and Coffino, P., *Proc. Natl. Acad. Sci. U.S.A.,* 74, 679, 1977.
37. Fujimoto, W. Y., Subak-Sharpe, J. H., and Seegmiller, J. W., *Proc. Natl. Acad. Sci. U.S.A.,* 68, 1516, 1971.
38. Garner, R. C. and Nutman, C. A., *Mutat. Res.,* 44, 9, 1977.
39. Garro, A. J., Guttenplan, J. B., and Milvy, P., *Mutat. Res.,* 38, 81, 1976.
40. Gilbert, W. and Muller-Hill, B., in *The Lactose Operon,* Beckwith, J. R. and Zipser, D., Eds., Cold Spring Harbor Laboratory, Cold Spring Harbor, N.Y., 1970, 93.
41. Gilbert, W. and Maxam, A., *Proc. Natl. Acad. Sci. U.S.A.,* 70, 3581, 1973.
42. Gilbert, W., Gralla, J., Majors, J., and Maxam, A., in *Proteinligand Interactions,* Walter de Gruyter, Berlin, 1975, 193.
43. Gilman, M. Z. and Thilly, W. G., *J. Therm. Biol.,* 2, 95, 1977.
44. Grant, E. L., Mitchell, R. H., West, P. R., Mazuch, L., and Ashwood-Smith, M. J., *Mutat Res.,* 40, 225, 1976.
45. Hince, T. A. and Neale, S., *Mutat. Res.,* 24, 383, 1974.
46. Hsie, A. W., Brimer, P. A., Mitchell, T. J., and Gosslee, D. G., *Som. Cell Gen.,* 1, 383, 1975.
47. Hsie, A. W., Brimer, P. A., Mitchell, T. J., and Gosslee, D. G., *Som. Cell Gen.,* 1, 247, 1975.
48. Huberman, E., Aspiras, L., Heidelberger, C., Grover, P. L., and Sims, P., *Proc. Natl. Acad. Sci. U.S.A.,* 68, 3195, 1971.
49. Huberman, E., Donovan, P. J., and DiPaolo, J. A., *J. Natl. Cancer Inst.,* 48, 837, 1972.
50. Huberman, E. and Heidelberger, C., *Mutat. Res.,* 14, 130, 1972.
51. Huberman, E. and Sachs, L., *Int. J. Cancer,* 13, 326, 1974.
52. Huberman, E. and Sachs, L., *Proc. Natl. Acad. Sci. U.S.A.,* 73, 188, 1976.
53. Inyer, V. N. and Szybalski, W., *Appl. Microbiol.,* 6, 23, 1958.
54. Ishii, Y. and Kondo, S., *Mutat. Res.,* 16, 13, 1972.
55. Ishii, Y. and Kondo, S., *Mutat. Res.,* 27, 27, 1975.
56. Isono, K. and Yourno, J., *Proc. Natl. Acad. Sci. U.S.A.,* 71, 1612, 1974.
57. Itiaba, K., Banfalvi, M., and Crawhall, J. C., *Biochem. Genet.,* 8, 149, 1973.
58. Jacob, F. and Monod, J., *J. Mol. Biol.,* 3, 318, 1961.
59. Jacobs, L. and DeMars, R., in *Handbook of Mutagenicity Test Procedures,* Kilby et al., Eds., Elsevier, Amsterdam, 1977, 193.
60. Jensen, K. A., Kirk, I., Kolmark, G., and Westergaard, M., *Cold Spring Harbor Symp. Quant. Biol.,* 16, 245, 1951.
61. Kao, F. T. and Puck, T. T., *J. Cell Physiol.,* 74, 245, 1969.
62. Kao, F. T. and Puck, T. T., *J. Cell Physiol.,* 78, 139, 1971.
63. Knapp, A. G. A. C. and Simons, J. W. I. M., *Mutat. Res.,* 30, 97, 1975.
64. Kondo, S., *Mutat. Res.,* 26, 235, 1974.
65. Krahn, D. and Heidelberger, C., *Mutat. Res.,* 46, 27, 1977.
66. Kuroki, T., Drevon, C., and Montesano, R., *Cancer Res.,* 37, 1044, 1977.
67. Maher, V. M. and Wessel, J. E., *Mutat. Res.,* 28, 277, 1975.
68. Mohn, G., *Mutat. Res.,* 20, 7, 1973.
69. McCann, J., Spingarn, N. E., Kobori, J., and Ames, B. N., *Proc. Natl. Acad. Sci. U.S.A.* 72, 979, 1975.
70. McCann, J., Choi, E., Yamasaki, E., and Ames, B. N., *Proc. Natl. Acad. Sci. U.S.A.* 72, 5135, 1975.
71. McCann, J. and Ames, B. N., *Proc. Natl. Acad. Sci. U.S.A.,* 73, 950, 1976.
72. Morrow, J., Prickett, M. S., Fritz, S., Vernick, D., and Deen, D., *Mutat. Res.,* 34, 481, 1976.
73. Nagao, M., Yahagi, T., Seino, Y., Sugimura, T., and Ito, N., *Mutat. Res.,* 42, 335, 1977.
74. Nagao, M., Honda, M., Seino, Y., Yahagi, T., and Sugimura, T., *Cancer Lett.,* 2, 221, 1977.
75. Oeschger, N. S. and Hartman, P. E., *J. Bacteriol.,* 101, 490, 1970.
76. Ohshima, Y., Horiuchi, T., and Yanagida, M., *J. Mol. Biol.,* 91, 515, 1975.
77. O'Neill, J. P., Brimer, P. A., Machanoff, R., Hirsch, G. P., and Hsie, A. W., *Mutat. Res.,* 45, 91, 1977.
78. O'Neill, J. P., Couch, D. B., Machanoff, R., San Sebastian, J. R., Brimer, P. A., and Hsie, A. W., *Mutat. Res.,* 45, 103, 1977.
79. Penman, B. W. and Thilly, W. G., *Som. Cell Genet.,* 2, 325, 1976.
80. Penman, B. W., Wong, M. V., and Thilly, W. G., *Life Sci.,* 19, 563, 1976.
81. Sato, H., Slesinski, R. S., and Littlefield, J. W., *Proc. Natl. Acad. Sci. U.S.A.,* 69, 1244, 1972.
82. Shapiro, N. I., Khalizev, A. E., Luss, E. V., Manuilova, E. S., Petrova, O. N., and Vanshaver, N. B., *Mutat. Res.,* 16, 89, 1972.
83. Shirasu, Y., Moriya, M., Kato, K., Furuhashi, A., and Kada, T., *Mutat. Res.,* 40, 19, 1976.
84. Silengo, L., Schlessinger, D., Mangiarotti, G., and Apirion, D., *Mutat. Res.,* 4, 701, 1967.
85. Skopek, T. R., Liber, H. L., Krolewski, J. J., and Thilly, W. G., *Proc. Natl. Acad. Sci. U.S.A.,* 75, 410, 1978.
86. Skopek, T. R., Liber, H. L., Kaden, D. A., and Thilly, W. G., *Proc. Natl. Acad. Sci. U.S.A.,* 75, 4470, 1978.
87. Stewart, C. R., *Genetics,* 59, 23, 1968.
88. Stoltz, D. R., Bandall, R. D., and Miller, C. T., *Mutat. Res.,* 40, 305, 1976.
89. Strauss, B. S., *Life Sci.,* 15, 1685, 1974.

90. Summers, W. P., *Mutat. Res.,* 20, 377, 1973.
91. Thilly, W. G. and Heidelberger, C., *Mutat. Res.,* 17, 287, 1973.
92. Thilly, W. G., DeLuca, J. G., Hoppe, H., IV, and Penman, B. W., *Chem. Biol. Int.,* 15, 33, 1976.
93. Van Zeeland, A. A. and Simons, J. W. I. M., *Mutat. Res.,* 35, 129, 1976.
94. Venitt, S. and Bushell, C. T., *Mutat. Res.,* 40, 309, 1976.
95. Whitfield, H. J., Jr., Martin, R. G., and Ames, B. N., *J. Mol. Biol.,* 21, 335, 1966.
96. Wong, J. J., Singh, R., and Hsieh, D. P. H., *Mutat. Res.,* 44, 449, 1977.
97. Yahagi, T., Matsushima, T., Nagao, M., Seino, Y., Sugimura, T., and Bryan, G. T., *Mutat. Res.,* 40, 9, 1976.
98. Yahagi, T., Nagao, M., Seino, Y., Matsushima, T., Sugimura, T., and Okada, M., *Mutat. Res.,* 48, 121, 1977.
99. Yanofsky, C., Ito, J., and Horn, V., *Cold Spring Harbor Symp. Quant. Biol.,* 31, 151, 1966.
100. Zampieri, A. and Greenberg, J., *Mutat. Res.,* 2, 5522, 1965.
101. Thilly, W. G., unpublished results.
102. Cox, R. and Masson, W. K., *Mutat. Res.,* 37, 125, 1976.
103. Thilly, W. G., *Toxicol. Environ. Health,* 2, 1343, 1977.
104. Poche, H., Vanshaver, N. B., Theille, M., Geissler, E., *Mutat. Res.,* 30, 83, 1975.
105. Termy, E. M., Venitt, S., and Brooks, P., *Mutat. Res.,* 19, 153, 1973.
106. Conney, A. H. and Burns, J. J., *Science,* 178, 4061, 1972.

UTILIZATION OF A QUANTITATIVE MAMMALIAN CELL MUTATION SYSTEM, CHO/HGPRT, IN EXPERIMENTAL MUTAGENESIS AND GENETIC TOXICOLOGY[a]

A. W. HSIE, D. B. COUCH,[b] J. P. O'NEILL,[b] J. R. SAN SEBASTIAN,[b] P. A. BRIMER, R. MACHANOFF, J. C. RIDDLE,[c] A. P. LI,[c] J. C. FUSCOE,[c] N. FORBES, and M. H. HSIE

TABLE OF CONTENTS

I. Introduction 40

II. Materials and Methods 41
    A. Development of a Stringent Selective System 41
    B. Coupling of the Microsome Activation System to the CHO/HGPRT Assay 41
    C. Host-Mediated CHO/HGPRT Assay 43

III. Results and Discussions 43
    A. Evidence for the Genetic Origin of Mutation Induction in the CHO/HGPRT System ...43
    B. Dose-Response Relationship for EMS-Mediated Mutation Induction and Cell Lethality ...44
    C. Apparent Dosimetry of EMS-Induced Mutagenesis 47
    D. Structure-Activity Relationship of Alkylating Agents and ICR Compounds 48
        1. Comparative Mutagenicity and Cytotoxicity of Alkylsulfate and Alkanesulfonate Derivitives ...48
        2. Mutagenicity and Cytotoxicity of Congeners of Two Classes of Nitroso Compounds ...48
        3. Mutagenicity of Heterocyclic Nitrogen Mustards (ICR Compounds) 48
    E. Mutagenicity of 79 Physical and Chemical Agents: Alkylating Chemicals, ICR Compounds, Metallic Compounds, Polycyclic Compounds, Nitrosamines, Miscellaneous Organic Chemicals, X-ray, UV Light, and other Nonionizing Agents ...48
        1. Alkylating Chemicals 49
        2. Heterocyclic Nitrogen Mustards — ICR Compounds (Total of Six) 49
        3. Metallic Compounds (Total of Eight) 49
        4. Polycyclic Hydrocarbons (Total of 27) 49
        5. Nitrosamines and Related Compounds (Total of Ten) 49
        6. Miscellaneous Compounds (Total of Ten) 50
        7. Physical Agents (Total of Seven) 50
    F. Preliminary Validation of the CHO/HGPRT Assay in Predicting Chemical Carcinogenicity ...50

IV. Concluding Remarks 50

References 53

---

[a] This paper is based on the material presented by the senior author at the Chemical Industry Institute of Toxicology (CIIT) Workshop on Strategies for Short-Term Testing for Mutagens/Carcinogens, Research Triangle Park, North Carolina, August 11 to 12, 1977. Research is sponsored jointly by the National Center for Toxicological Research, the Environmental Protection Agency, and the Department of Energy under contract with the Union Carbide Corporation.

[b] Postdoctoral Investigator, Carcinogenesis Training Grant No. CA 05296 from the National Cancer Institute.

[c] Predoctoral Fellow: JCR, Predoctoral Training Grant No. GM 01974 from the National Institute of General Medical Sciences, National Institutes of Health; APL, University of Tennessee Research Assistantship; JCF, Carcinogenesis Training Grant No. CA 09104 from the National Cancer Institute.

I. INTRODUCTION

It has been recognized recently that studies of mutagenesis in prokaryotes may not reveal some fundamental mechanisms of mutagenesis in mammals, because mammals differ from prokaryotes in their level of organization and repair of DNA, mechanisms of metabolism of chemicals, and other related functions. Since the observation that treatment of mammalian somatic cells with conventional mutagens, such as ethyl methanesulfonate (EMS) and N-methyl-N'-nitro-N-nitrosoguanidine (MNNG), causes an increase of cell variants that differ from the parental cells in either nutritional requirements[4,21,35] or drug sensitivity,[4] there has been much interest in utilizing a quantitative mutation system for studying mechanisms underlying the process of mammalian mutation and, additionally, for assessing the genetic hazard of environmental agents to the human population. Several mammalian cell mutation systems have been developed for such purposes.[5]

Since the discovery that mammalian cells resistant to the purine analogue 8-azaguanine (AG) — or, more recently, 6-thioguanine (TG) — can be selected after treatment with chemical mutagens, a variety of mutagenesis studies have utilized resistance to purine analogues as a genetic marker.[4,5]

The selection of mutation induction to purine analogue resistance is based on the fact that the wild-type cells containing hypoxanthine-guanine phosphoribosyl transferase (HGPRT) activity are capable of converting the analogue to toxic metabolites, leading to cell death; the presumptive mutants, by virtue of the loss of HGPRT activity, are incapable of catalyzing this detrimental metabolism and, hence, escape the lethal effect of the purine analogue. Because of the relative ease of obtaining, scoring, and characterizing resistance to purine analogues and of studying the forward as well as the reverse mutation (resistance to aminopterin), this drug-resistance marker has been widely used in mammalian cell mutagenesis studies. The HGPRT gene is on the functionally monosomic X chromosome in diploid mammalian somatic cells. Only one X chromosome is functional genetically: male cells have one functional X chromosome because the Y chromosome does not contain genes known to be on the X chromosome, and female cells also have only one functional X chromosome because the other X chromosome has been inactivated during early embryogenesis, according to the X-inactivation mechanism proposed originally by Russell[37] and Lyon.[24] Therefore, one would expect that mutation of the diploid cells at the HGPRT locus would be similar to that of the bacteria.[34] This attribute greatly facilitates studies of induction of mutation and the subsequent analyses of mutagenesis.

Despite the remarkable progress in somatic cell mutagenesis studies over the years,[5] there remain numerous conflicting reports concerning conditions affecting mutagenesis, including even the most fundamental problem of the genetic basis of mutation induction to purine analogues. To clarify aspects of these problems, we have redefined conditions necessary for the selection of purine analogue resistance, utilizing Chinese hamster ovary (CHO), cells clone K_1-BH_4, and TG to measure mutation induction at the HGPRT locus; we refer to the assay as the CHO/HGPRT system or assay.

CHO cells were chosen for the study because they are perhaps the best-characterized mammalian cells genetically and are suitable for studies in mutagenesis.[22,34] In addition, they exhibit high cloning efficiency, achieving nearly 100% under normal growth conditions, and are capable of growing in a relatively well-defined medium on a glass or plastic substratum or in suspension with a short population doubling time of 12 to 13 hr. They have a stable, easily recognizable karyotype of 20 or 21 chromosomes (depending on the subclone), which appears to have undergone extensive chromosomal rearrangement.[9] This may have resulted in many hemizygous regions, since numerous phenotypically recessive mutations have been isolated (Table 1).[40]

Purine analogues AG and TG are known to differ in the mechanism by which they cause growth inhibition and cellular lethality to mammalian cells.[36] The choice of TG over AG as a selective agent has been prompted by reports that TG-resistant variants result from an alteration or loss of the enzyme HGPRT,[3,39] implying that such mutations occur at the structural gene that confers HGPRT activity. In addition, we have found that AG at modest concentrations lacks the stringency in selecting variants

TABLE 1

Characteristics of CHO Cells

1. Exhibit a stable karyotype over 20 years with a modal chromosome number of 20, which has a distinctly recognizable morphology
2. Have a colony-forming capacity of nearly 100% in a defined growth medium
3. Grow well in either monolayer or suspension with a relatively short population doubling time of 12 to 14 hr
4. Are genetically and biochemically well characterized, with many genetic markers available, including auxotrophy, drug resistance, temperature sensitivity, etc.
5. Respond well to various synchronization methods, including the mitotic detachment procedure that facilitates cell cycle study
6. Are useful in somatic cell hydridization experiments because they readily hybridize with different cell types, including human cells; when the CHO-human cell hybrid is formed there is subsequent rapid, preferential loss of human chromosome, which facilitates the assignment of marker genes to specific chromosomes or linkage groups in the human karyotype
7. Quantitatively respond to various physical and chemical mutagens and carcinogens with high sensitivity
8. Adapt to mutation induction either through coupling with a microsome activation system or through host (mouse) mediation
9. Are capable of monitoring induced mutation to multiple gene markers, chromosome aberration, and sister chromatid exchange in the same mutagen-treated cell culture

affected at the HGPRT locus and at higher levels exhibits general cytotoxicity leading to lethal effects to both the wild-type and the HGPRT-deficient variants.[12]

II. MATERIALS AND METHODS

A. Development of a Stringent Selective System

The CHO/HGPRT system has been defined in terms of medium, TG concentration, optimal cell density for selection (and, hence, recovery of the presumptive mutants), and expression time for the mutant phenotype.[13,28] Routinely, approximately 1×10^6 cells were exposed to a mutagen for a fixed period of time, usually 16 hr, which covers slightly over one population doubling time. After the mutagen was removed, the treated cells were allowed to express the "mutant phenotype" in F12 medium for 7 to 9 days, at which time mutation induction reached a maximum that was maintained thereafter (as long as 35 days examined) for several agents (EMS, MNNG, ICR-191, X-ray, and UV), irrespective of concentration or intensity of the mutagen (Figure 1). Routine subculture was performed at 2-day intervals during the expression period, and at the end of this time the cells were plated for selection in hypoxanthine-free F12 medium containing $1.7\mu/m\ell$ (10 μM) of TG at a density of 2.0×10^5 cells per 100-mm plastic dishes (Corning® or Falcon®), which permits 100% mutant recovery in reconstruction experiments (Figure 2). We find the use of dialyzed serum particularly important, presumably due to potential competition between hypoxanthine and TG for catalysis by HGPRT (Figure 3) and for transport into the cells.[1] After 7 to 8 days in the selective medium, the drug-resistant colonies developed; they were then fixed, stained, and counted. Such a protocol permits the maximum yield of variants selected at the HGPRT locus by various physical and chemical agents.[7,13,14,16,18,28,29,31] Mutation frequency was calculated based on the number of drug-resistant colonies per survivor at the end of the expression period. Weighted least-squares regression analysis was employed to determine the shape of the curves for mutation induction and cell survival.[13]

B. Coupling of the Microsome Activation System to the CHO/HGPRT Assay

Promutagens or procarcinogens are inactive by themselves, but are converted to "active principles" through biotransformation. The CHO/HGPRT assay can be coupled with the so-called S_9 fraction derived from rat-liver microsome-activation preparations,[29] as first described by Krahn and Heidelberger for V79 cells[23] for determining mutagenicity of procarcinogens, such a benzo(a)pyrene, benz-(a)anthracene, and dimethylnitrosamine. In our studies, the microsome fraction prepared from

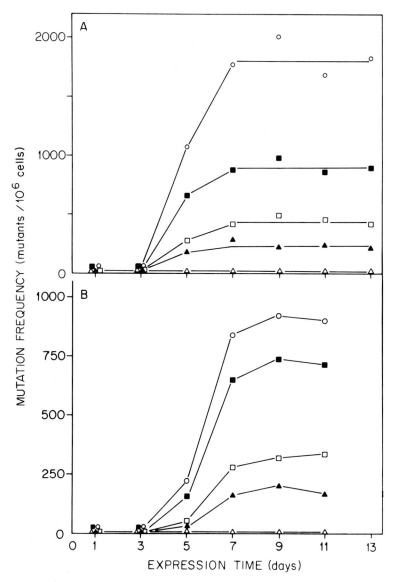

FIGURE 1. Expression time of HGPRT-(TG-resistant) mutant phenotype. CHO cells (clone K_1-BH_4) were plated at 5×10^5 cells per 100-mm plate in Ham's F12 medium containing 5% dialyzed fetal calf serum and were incubated at 37°C in 5% CO_2 in air in a 100% humidified incubator. After 24 hr (cell no. = 1 to 1.5×10^6 cells per plate) either (A) EMS or (B) ICR-191 was added (time = 0), and the cultures were incubated for 16 hr. The cultures were washed three times with saline, the cells were removed by trypsinization, and approximately 0.5 to 1.0×10^6 cells were subcultured every 48 hr. Mutant selection was performed at day 1 and at 2-day intervals thereafter. For selection of the mutant phenotype, cells were plated at 2×10^5 cells per 100-mm plate (5 plates = 10^6 cells) in hypoxanthine-free F12 medium containing 5% dialyzed fetal calf serum and 10 μM TG. For cloning efficiency, 200 to 1000 cells were plated in triplicate in the same medium without TG. After 7 days of growth, colonies were fixed, stained, and counted. Mutation frequency was calculated as the number of mutant colonies per 10^6 cells (corrected for the cloning efficiency at the time of selection). (A) Cells treated with EMS concentrations ($\mu g/m\ell$) at: 0 (△), 50 (▲), 100 (□), 200 (■), or 400 (○) were determined as described. (B) Cells treated with ICR-191 concentrations ($\mu g/m\ell$) at: 0 (△), 0.005 (▲), 0.1 (□), 0.25 (■), and 0.5 (○) were determined as described. The average spontaneous mutation frequency was 2 to 3.6×10^{-6}. (From O'Neill, J. P., Fuscoe, J. N., and Hsie, A. W., *Cancer Res.,* 38, 506, 1978; O'Neill, J. P. and Hsie, A. W., *Nature(London),* 269, 815, 1977. With permission.)

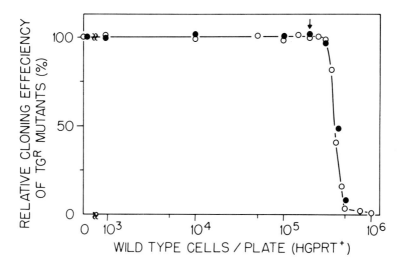

FIGURE 2. Effect of HGPRT⁺ cell number on recovery of HGPRT⁻ cells in TG medium. Relative cloning efficiency of HGPRT⁻ cells is plotted vs. the number of HGPRT⁺ (wild-type) cells. HGPRT⁻ cells (200) were plated in the presence of varying numbers of wild-type cells in TG medium on 100-mm plates. After 7 days the colonies were fixed, stained, and counted. The HGPRT⁻ cells are MNNG-A (O, absolute cloning efficiency = 0.88; HGPRT activity [% of wild type] = 9.4) and EMS 50-1 (●, cloning efficiency = 0.94; HGPRTase activity [% of wild type] = 5.4). The arrow marks a cell number of 2×10^5 cells per 100-mm plate. (From O'Neill, J. P., Brimer, P. A., Machanoff, R., Hirsch, G. P., and Hsie, A. W., *Mutat. Res.*, 45, 91, 1977. With permission from Elsevier/North-Holland Biomedical Press.)

Aroclor 1254-induced, male Spraque-Dawley rat livers was the only activation source employed. Numerous factors that might affect metabolic activation both qualitatively and quantitatively, such as differences in organ, sex, animal species, and inducer, await further studies.

C. Host-Mediated CHO/HGPRT Assay

An alternative method of achieving metabolic activation is the use of a host-mediated assay.[10] Comparison of mutagenic activity of a compound in the host-mediated assay with that in direct tests can indicate whether activation or detoxification has occurred. Nude mice, or their phenotypically normal littermates, and BALB/c mice have been used as hosts for CHO/HGPRT assays to determine the mutagenicity of dimethylnitrosamine, benzo(a)pyrene, and EMS.[17,18]

III. RESULTS AND DISCUSSIONS

A. Evidence for the Genetic Origin of Mutation Induction in the CHO/HGPRT System

Employing the assay protocol described in Section II and outlined in Table 2, we have found that the spontaneous mutation frequency lies in the range of 1 to 5×10^{-6} mutant per cell, as found in approximately 400 experiments over the past 4 years. Various physical and chemical agents are capable of inducing TG resistance. Among all chemical mutagens examined, mutation induction occurs as a linear function of the concentration.[7,12-14,16,18,28-31] For example, mutation frequency increases approximately linearly with EMS concentration in a near-diploid CHO cell line, conforming to the expectation that mutation induction occurs in the gene localized at the functionally monosomic X chromosome. However, in the tetraploid CHO cells EMS does not induce an appreciable number of mutations, even at very high concentrations. The virtual absence of EMS-induced mutation to TG resistance in tetraploid cells is expected, as the theoretically predicted mutation frequency would be the spontaneous frequency (which is estimated to be 10^{-12} mutants per survivor) plus the square of the induced mutation frequency at the corresponding EMS concentration in the diploid cells (which is no greater than 5×10^{-6} mutants per survi-

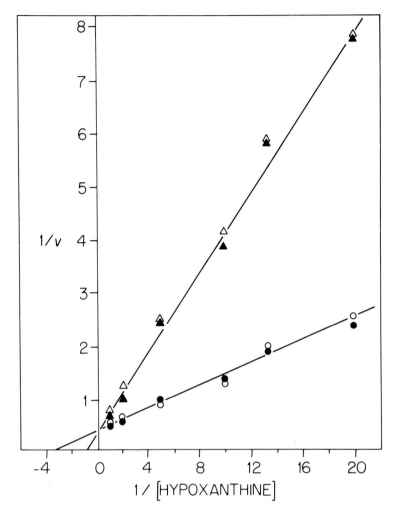

FIGURE 3. Lineweaver-Burke graph of CHO cell HGPRT activity. The reaction contained hypoxanthine over a concentration range of 0.5 to 10 μM, in the absence (△, ▲) or presence (○, ●) of 10 μM TG with 5 μg (open symbols) or 10 μg (closed symbols) cell extract protein. Velocity (v) is expressed in nanomoles per minute per milligram protein. From the intercepts, the apparent K_m for hypoxanthine is 2.5 μM and the apparent K_i for TG is 2.6 μM. (From O'Neill, J. P., Brimer, P. A., Machanoff, R., Hirsch, G. P., and Hsie, A. W., Mutat. Res., 45, 91, 1977. With permission from Elsevier/North-Holland Biomedical Press.)

vor).[12] We have been unable to detect any spontaneous reversion with 13 TG-resistant mutants, all of which contain low, yet detectable, HGPRT activity. Over 98% of the presumptive mutants isolated either from spontaneous mutation or as a result of mutation induction are sensitive to aminopterin, incorporate hypoxanthine at reduced rates, and have less that 5% HGPRT activity.[12,28] Studies in progress have also shown that mutants containing temperature-sensitive HGPRT activity can be selected at a relatively high frequency, suggesting that mutation resides in the HGPRT structural gene.

The CHO/HGPRT system appears to fulfill the criteria for a specific gene locus mutational assay (Table 3) and should be valuable in studying mechanisms of mammalian cell mutagenesis and in determining the mutagenicity of various physical and chemical agents.

B. Dose-Response Relationship for EMS-Mediated Mutation Induction and Cell Lethality

The first quantitative mutagenesis experiment performed was the dose-response of EMS. We found that EMS-induced mutation frequency to TG resistance is a linear function over a large range of mutagen concentrations

TABLE 2

CHO/HGPRT Mutation Assay

I. Enzyme System

II. Mutation Induction and Selection for Variants and Revertants

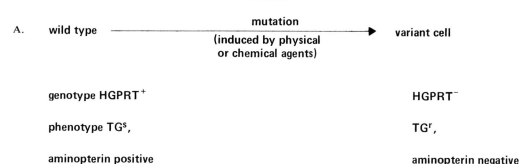

B. Variant selection is based on resistance to TG
C. Selection of revertants is based on growth in presence of aminopterin

III. Characterization of TG^r Variants
  A. Direct enzyme for conversion of (^{3}H)hypoxanthine to (^{3}H)IMP
  B. Cellular incorporation of (^{3}H)hypoxanthine into cellular macromolecules, as revealed by either direct radio-activity measurement or autoradiographic determination
  C. Sensitivity of clonal growth to aminopterin (10 μM) in medium F12FCM5, which contains hypoxanthine (30 μM), glycine (100 μM), and thymidine (3 μM)

TABLE 3

CHO/HGPRT Mutation Assay: Genetic Basis of Mutation at HGPRT Locus in TG-Resistance Selection

I. Spontaneous mutation frequency at 1 to 5×10^{-6} mutant per cell

II. Mutation induction by physical and chemical agents with linear dose-response relationship

III. Frequency of spontaneous reversion less than 10^{-7} reversion per cell

IV. Failure to induce mutation in near-tetraploid cell lines

V. Altered HGPRT activity in mutants
  A. 1170/1189 (98.4%) mutant colonies are aminopterin sensitive
  B. 121/122 (99.2%) mutant colonies show reduced hypoxanthine incorporation by autoradiography studies
  C. 81/83 (97.6%) isolated mutant clones show reduced HGPRT enzyme activity

From O'Neill, J. P., Brimer, P.A., Machanoff, R., Hirsch, G. P., and Hsie, A. W., *Mutat. Res.*, 45, 91, 1977. With permission from Elsevier/North-Holand Biomedical Press.

when cells are treated for a fixed period of 16 hr. The mutation-induction line can be adequately described by $f(x) = (8.73 + 3.45x) \times 10^{-6}$, where $f(x)$ is the fitted mutation frequency and x is the EMS concentration. These EMS concentrations cover a wide range of the survival curve, including the shoulder region (0 to 100 $\mu g/m\ell$ of EMS), where no appreciable loss of cellular viability occurs, and the straight portion (100 to 800 $\mu g/m\ell$ of EMS), where cell killing increases nearly exponentially with increasing mutagen concentrations (Figure 4).[13]

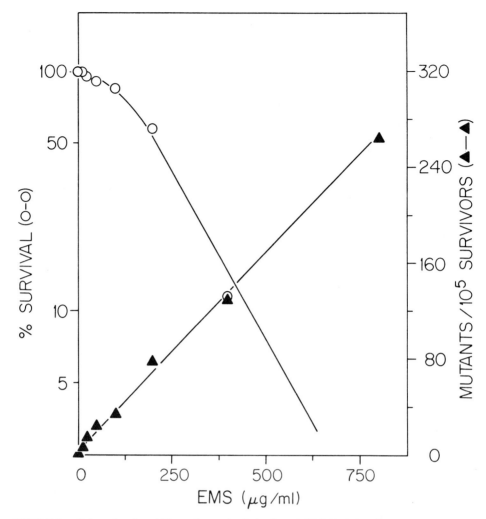

FIGURE 4. Colony-forming ability and mutation induction of CHO-K_1-BH_4 cells treated with various EMS concentrations for 16 hr under the standard conditions described in the text. The survival curve (O) was constructed from the pooled data of two separate experiments by fitting the data to the "multi-target model"

$$R_i = 1 - (1 - e^{Bx_i})^N$$

where B and N are fitted coefficients. The line (▲) was constructed using weighted least-square analysis from the pooled data of two separate experiments. The straight-line fit of mutation frequency, $f(x)$, in mutants per 10^6 survivors as a function of EMS concentration, x (in μg/mℓ is $f(x) = (8.73 + 3.45x) \times 10^{-6}$. The 95% confidence limits for the slope are 3.02×10^{-6}, 3.88×10^{-6}. (From Hsie, A. W., Brimer, P. A., Mitchell, T. J., and Goslee, D. G., *Somat. Cell Genet.*, 1, 247, 1975. With permission.)

Apparently, no threshold effect is exhibited for mutation induction by EMS. It is worth noting that mutation frequency reaches approximately 350×10^{-6} at 100 μg/mℓ of EMS (from the spontaneous frequency at 9×10^{-6}), while no significant cell killing occurs. This experiment demonstrates that mutation induction could occur without significant loss of cell survival and that it is a more sensitive parameter of genetic toxicity than cell lethality, as far as agents such as EMS are concerned.

Later experiments show that many physical and chemical agents exhibit an "EMS-type" curve of interrelationships of cell survival and mutation induction.[7,14,16,18,29,31] However, there are agents, such as MNNG, which do not exhibit an appreciable shoulder region in the survival curve, and mutation induction always occurs concomitantly with the loss of cell survival.[7,28,30]

C. Apparent Dosimetry of EMS-Induced Mutagenesis

To investigate whether EMS-induced mutagenesis can be quantified further, cells were treated with several concentrations of EMS for intervals of 2 to 24 hr. A linear increase in mutation frquency was found with EMS concentrations of 50 to 400 µg/mℓ for incubation times of up to 12 to 14 hr. However, cell survival decreased exponentially with time over the entire 24-hr period. This difference in the time course of cellular lethality vs. mutagenicity might be due to the formation of toxic, nonmutagenic breakdown products in the medium with longer incubation times or might reflect a difference in the mode of action of EMS in these two biological effects. However, the increases in both cell lethality and induced mutation with incubation times of up to 12 hr allow a study of the apparent dosimetry of this chemical mutagen. For better defined dosimetry studies, cultures were incubated with varying concentrations of EMS (0.05 to 3.2 mg/mℓ) for 2 to 12 hr. Here we define the apparent dose of EMS to be the product of the EMS concentration and the treatment time. The response to mutagen treatment (dose) should be the same for different combinations of concentration multiplied by duration of treatment which yield the same product. As shown in Figure 5, the decrease in cell survival was exponential and the increase in mutation frequency was linear with dose. From these studies the mutagenic potential of EMS can be described as 310×10^{-6} mutants (cell mg mℓ$^{-1}$ hr)$^{-1}$.

These results demonstrate that the mutagenicity of EMS in mammalian cells follows a linear dosimetry relationship and that a quantitative study of chemical mutagenesis in mammalian cells can be performed.

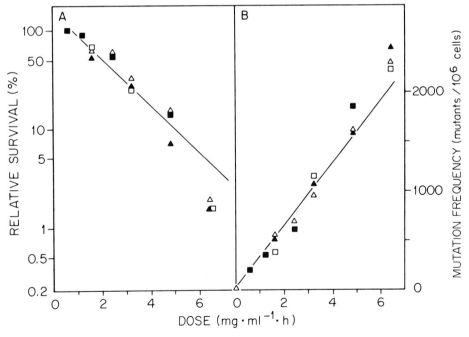

FIGURE 5. Apparent dosimetry for EMS cellular lethality (A) and induced mutation frequency (B). Cells were treated with various concentrations (mg/mℓ) of EMS for 2 (△), 4 (▲), 8 (□), or 12 hr (■). After treatment the cultures were washed, and single-cell survival was determined after a total time of 24 hr. Cells were plated for selection on day 9 of expression, as in Figure 1. Spontaneous mutation frequency = 2.3×10^{-6}. In (A) the linear fit of the survival curve was found over 1.6 to 6.4 mg (of EMS)/mℓ$^{-1}$ hr through fitting to the "multitarget model". The extrapolation number, N, is 1.74, with 95% confidence limits of 1.18, 2.58; the slope, B, is -0.248, with 95% confidence limits of -0.303, -0.188. In (B) the straight-line fit for the pooled data on mutation induction over the dose range 0.64 to 6.4 mg (of EMS)/mℓ per hour is $f(x) = (5.58 \times 310.77x) \times 10^{-6}$. (From O'Neill, J. P. and Hsie, A. W., *Nature (London)*, 269, 815, 1977. With permission from Macmillan Journals Ltd.)

D. Structure-Activity Relationship of Alkylating Agents and ICR Compounds

The quantitative nature of mutation induction by EMS[12,13,31] and ICR-191[29] suggests the applicability of the CHO/HGPRT system for comparing the mutagenicity and cytotoxicity of various direct-acting alkylating agents and ICR compounds.

1. Comparative Mutagenicity and Cytotoxicity of Alkylsulfate and Alkanesulfonate Derivatives

Dose-response relationships of cell killing and mutation induction of two alkylsulfates (dimethylsulfate [DMS] and diethylsulfate [DES]) and three alkyl alkanesulfonates (methyl methanesulfonate [MMS], EMS, and isopropyl methanesulfonate [iPMS]) have been studied, and certain features of these relationships are summarized here (Table 4).[7,8] Under the experimental conditions employed, cytotoxicity decreased with the size of the alkyl group: DMS>DES; MMS>EMS>iPMS. All agents produced linear dose-response of mutation induction: based on mutants induced per unit mutagen concentration, DMS>DES; MMS>EMS>iPMS. However, when comparisons were made at 10% survival, mutagenic potency was DES>DMS; EMS>iPMS>MMS.

2. Mutagenicity and Cytotoxicity of Congeners of Two Classes of Nitroso Compounds

Using the CHO/HGPRT system, we have compared the cytotoxicity and mutagenicity of two nitrosamidines (MNNG and N-ethyl-N'-nitro-N-nitrosoguanidine [ENNG]) and three nitrosamides (N-methyl-N-nitrosourea [MNU], N-ethyl-N-nitrosourea [ENU], and N-butyl-N-nitrosourea [BNU]) differing in the nature of their alkylating group. Based on 10% survival data, the following order of cytotoxicity was observed: MNNG>ENNG>MNU>ENU>BNU. This is the same order of potency as observed for mutation induction per unit concentration of mutagen (Table 5).[6,7]

3. Mutagenicity of Heterocyclic Nitrogen Mustards (ICR Compounds)

The six compounds chosen (ICR-191,-170,-292,-372,-191-OH, and -170-OH) contain structural similarities and differences which allow the roles of both the heterocyclic nucleus and the alkylating side chain in the mutagenic activity to be determined. The first four contain a single 2-chloroethyl group (nitrogen half-mustard) on the side chain and are mutagenic, with the tertiary amine types (170 and 292) three to five times more mutagenic than the secondary amine types (191 and 372). The two remaining compounds (191-OH and 170-OH) are not mutagenic, indicating that the 2-chloroethyl group is needed for mutation induction.[29,30]

E. Mutagenicity of 79 Physical and Chemical Agents: Alkylating Chemicals, ICR Compounds, Metallic Compounds, Polycyclic Hydrocarbons, Nitrosamines, Miscellaneous Organic Chemicals, X-ray, UV Light, and Other Nonionizing Agents

So far, mutagenicity has been studied in 79 physical and chemical agents, 42 of which have

TABLE 4

Cytotoxicity and Mutagenicity of Alkyl Sulfates and Alkanesulfonates

Compound	Concentration (μM) to produce 10% survival	Mutation frequency (mutants/10^6 survivors)	
		At 10% survival	Per μM mutagen
DMS	89	92	1.0
DES	2760	780	0.3
MMS	95	140	1.5
EMS	3700	1550	0.4
iPMS	4540	435	0.1

From Couch, D. B., Forbes, N. L., and Hsie, A. W., *Mutat. Res.*, 51, 217, 1978. With permission from Elsevier/North-Holland Biomedical Press.

documented carcinogenicity or noncarcinogenicity.[19,41] Included in this study were several classes of chemicals such as polycyclic hydrocarbons and nitrosamines, which require coupling of metabolic activation systems to be mutagenic and cytotoxic. As a means of validating the CHO/HGPRT assay as a prescreen for carcinogens, several noncarcinogenic structural analogues of carcinogens were also included. Most of the mutagenicity studies were conducted at cell survivals ranging from 100 to 10%, and some to 1%. Several points can be made.

1. Alkylating Agents and Related Compounds (Total of 11)

Included in the mutagenicity study are two alkylsulfates, three alkanesulfonates, two nitrosamidines, three nitrosamides, and N-methyl-N'-nitroguanidine. All ten alkylating agents (Tables 4 and 5) are carcinogens and are highly mutagenic. The fact that the carcinogen MNNG (N-methyl-N'-nitro-N-nitrosoguanidine) is mutagenic and the structural analogue N-methyl-N'-nitroguanidine is neither carcinogenic nor mutagenic implies that nitrosation is required for both mutagenicity and carcinogenicity.

2. Heterocyclic Nitrogen Mustards — ICR Compounds (Total of Six)

Two ICR compounds (170 and 292) are carcinogens[32] and are highly mutagenic. Noncarcinogenic ICR-191[32] is an active mutagen in the CHO/HGPRT system and was shown to be a powerful frameshift mutagen in the *Salmonella* mutation system.[2,25,26]

3. Metallic Compounds (Total of Eight)

The mutagenicity of four carcinogenic metals (Ni, Be, Cd, and Co salts) can be demonstrated in this system. The CHO/HGPRT system appears to be useful for studying mutagenesis of metallic compounds whose mutagenicity was not detectable in the *Salmonella* system.[2,25,26]

4. Polycyclic Hydrocarbons (Total of 27)

We studied the mutagenicity of benzo(a)pyrene [B(a)P] and 19 metabolites, including 11 phenols, 3 epoxides, 3 diols, and 2 diolepoxides. For comparison, benzo (e)pyrene and pyrene were added to this study. Also included were benz(a)anthracene (BA) and four related compounds (7,12-dimethyl BA, anthracene, and two phenolic derivatives of BA). The carcinogenic polycyclic hydrocarbons (B(a)P, BA, and 7,12-dimethyl BA) require metabolic activation to be mutagenic. The noncarcinogenic polycyclic hydrocarbons pyrene and anthracene are nonmutagenic even with metabolic activation. Since CHO cells cannot activate procarcinogens such as B(a)P, these cells appear to be most useful in screening for the mutagenicity of metabolites such as the 4,5-epoxide of B(a)P.[11]

5. Nitrosamines and Related Compounds (Total of Ten)

All four carcinogenic nitrosamines (dimethylnitrosamine, diethylnitrosamine, nitrosopyrrolidine, and 2-methyl nitropiperidine) also require metabolic activation to be mutagenic, while noncarcinogenic ones (2,5-dimethyl nitrosopyrrolidine and 2,5-dimethylpiperidine) are not mutagenic in this system. The requirement

TABLE 5

Cytotoxicity and Mutagenicity of Nitrosoguanidines and Nitrosoureas[6,7,28]

Compound	Concentration (μM) to produce 10% survival	Mutation frequency (mutants/10^6 survivors)	
		At 10% survival	Per μM mutagen
MNNG	0.34	200	590
ENNG	6.1	522	86
MNU	86	243	3
ENU	1200	550	0.5
BNU	2540	200	0.08

From Couch, D. B. and Hsie, A. W., *Mutat. Res.*, 57, 209, 1978. With permission from Elsevier/North-Holland Biomedical Press.

of nitrosation for both mutagenicity and carcinogenicity can be demonstrated from the observations that carcinogenic dimethylnitrosamine is mutagenic with activation, while dimethylamine is not mutagenic irrespective of activation. Formaldehyde, a metabolite of dimethylnitrosamine, is carcinogenic in some systems but not in others;[27] its mutagenicity is not evident in this system. Neither nitrate nor nitrite appears to be mutagenic.[38]

6. Miscellaneous Compounds (Total of Ten)

Three commonly used organic solvents (acetone, dimethylsulfoxide, and ethanol) are noncarcinogens and do not appear to be mutagenic. All four metabolic inhibitors (cytosine arabinoside, hydroxyurea, caffeine, and cycloheximide) are nonmutagenic in a preliminary study without the addition of S_9. Hydrazine and hycanthone appear to be direct-acting mutagens because they are mutagenic without S_9. $N^6,O^{2'}$-dibutyryl adenosine 3':5'-phosphate, an analogue of adenosine 3':5'-phosphate, an important effector of growth and differentiation in many biological systems, is not mutagenic.[15]

7. Physical Agents (Total of Seven)

The mutagenicity of both ionizing radiation such as X-ray[29] and nonionizing physical agents such as UV light[14] can be demonstrated. Fluorescent white, black, and blue lights were slightly lethal and mutagenic.[18] Sunlamp light was highly lethal and mutagenic, exhibiting the biological effects within 15 sec of exposure under conditions recommended by the manufacturer for human use.[16] Lethal and mutagenic effects were observed after 5 min of sunlight exposure (Figure 6); responses varied with hourly and daily variations in solar radiation (Table 6).[16] In view of man's constant exposure to and use of various light sources, demonstration of their genetic toxicity indicates that daily exposure to these light sources, especially sunlight, should be minimized. The demonstration that the CHO/HGPRT system is capable of quantifying the lethal and mutagenic effect of sunlight recommends it as a model mammalian cell system for studies of the genetic toxicology of sunlight per se and of the interactive effects between sunlight and other physical and chemical agents, leading ultimately to a better understanding of the effects of sunlight on man.

F. Preliminary Validation of the CHO/HGPRT Assay in Predicting Chemical Carcinogenicity

Mutagenicity in the CHO/HGPRT assay of 42 chemicals (whose carcinogenicity or noncarcinogenicity has been demonstrated[19,41] correlated well (40/42 [95.24%]) with the reported carcinogenicity in the animal tests (Table 7). The existence of such a good correlation between mutagenicity and carcinogenicity speaks favorably for the utility of this assay in prescreening the carcinogenicity of chemical and physical agents. However, this result should be viewed with caution since only limited classes of chemicals have been tested so far.

A possible false negative was shown for formaldehyde, which has been shown to be either carcinogenic or noncarcinogenic depending on the means of exposure to the test animals.[27] Apparently, a false positive was ICR-191, a well-known potent mutagen in Salmonella,[2,25,26] which has been shown to be noncarcinogenic in a recent study.[32]

IV. CONCLUDING REMARKS

We feel that we have now come close to optimizing conditions necessary for quantifying mutation induction to TG resistance in the CHO/HGPRT system, which selects for >98% mutants deficient in the activity of HGPRT specified by a gene on the X chromosome, in a near-diploid CHO cell line. In all chemical mutagens studied in the CHO/HGPRT system, mutation induction increases linearly as a function of the concentration without an apparent threshold effect. In contrast to the linear dose-response of EMS-induced mutation in the diploid cells, EMS failed to induce mutations in near-tetraploid CHO cells, consistent with the notion that induced mutation occurs at the gene or chromosomal level. As an example of dosimetry studies, EMS produced 310×10^{-6} mutants (cell mg mℓ^{-1} hr)$^{-1}$. The sensitive and quantitative nature of this assay has been utilized to study the structure-activity (mutagenicity) relationship of chemicals.

In addition to being capable of determining mutagenicity of direct-acting agents, the system can detect mutagenicity of procarcinogens by coupling with a standard microsome activation system or through a host (BALB/c mouse)-mediated assay. Thus far, mutagenicity has been

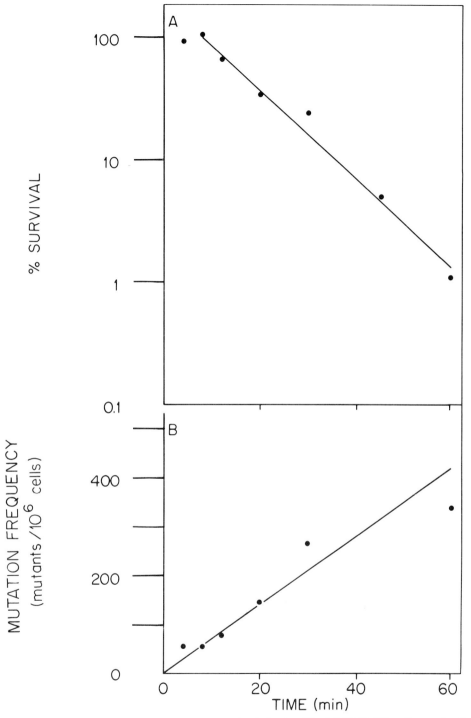

FIGURE 6. Colony-forming ability (A) and mutation induction (B) of CHO-K_1-BH_4 cells as a function of sunlight exposure time on July 17, 1975. For analysis of the survival curve, the linear fit was determined over the exposure-time range of 12 to 60 min through fitting to the "multitarget model". Th extrapolation number, N, is 1.94, with 95% confidence limits of 1.15 and 3.27; the lope, B, is $-3,56 \times 10^{-2}$, with 95% confidence limits of -4.33×10^{-2} and -2.86×10^{-2}. For analysis of the mutation-induction curve, the linear fit was determined for an exposure period of 0 to 60 min. The straight-line fit is $f(x) = (1.29 + 6.99x) \times 10^{-6}$. The 95% confidence limits for the slope are 5.26×10^{-6}, 8.73×10^{-6}. Nearly identical curves for cell survival and mutation induction were found in experiments performed on July 18, 1975, with weather conditions (furnished by the U.S. Department of Commerce Weather Bureau at Oak Ridge) almost identical to those of July 17, 1975. (From Hsie, A. W., Li, A. P., and Machanoff, R., *Mutat. Res.*, 45, 333, 1977. With permission from Elsevier/North-Holland Biomedical Press.)

TABLE 6

Lethality and Mutagenicity of Sunlight at Different Times of Day

Time	Ambient temperature (°C)	Solar radiation (g cal/cm² min)	Loss of relative survival (%)	Observed mutation frequency (TG mutants/10⁶ cells)
6:00 a.m. (control)	17.2	3.8	0	5
6:45 a.m.	17.2	10.6	34	5
9:00 a.m.	21.7	55.1	55	12
12:30 p.m.	28.9	75.0	65	147
4:30 p.m.	30.6	37.1	40	32
6:30 p.m.	26.1	9.8	25	10

Note: All experiments were performed on July 28, 1975, with an exposure time of 10 min, except the untreated control.

From Hsie, A. W., Li, A. P., and Machanoff, R., *Mutat. Res.*, 45, 333, 1977. With permission from Elsevier/North-Holland Biomedical Press.

TABLE 7

Correlation of Mutagenicity[a] in the CHO/HGPRT Assay with Reported Carcinogenicity[b] in Animal Tests

Agent[c]	Concurrence[d]	False negatives[e]	False positives[f]
Alkylating agents and relatives	11/11 (100%)	0	0
ICR compounds	2/3 (66.7%)	0	1/3 (33.3%)
Metallic compounds	4/4 (100%)	0	0
Polycyclic hydrocarbons	6/6 (100%)	0	0
Nitrosamines and related compounds	9/10 (90%)	1/10 (10%)	0
Miscellaneous agents	5/5 (100%)	0	0
Physical agents	3/3 (100%)	0	0
All agents	40/42 (95.24%)	1/42 (2.38%)	1/42 (2.38%)

[a] Agents studied are found to be either mutagenic, regardless of "mutagenic potency", or nonmutagenic. Since only limited experiments were performed on a small sample of each derivative of B(a)P, these results should be viewed as preliminary in nature; more studies on these compounds are under way. The mutagenicity is assayed either directly or coupled with a metabolic activation system in vitro or in vivo. In the S₉-coupled assay the microsome used was prepared from Aroclor 1254-induced, male Sprague-Dawley rat livers. The effects of either other inducers or conditions of the activation system have not been investigated extensively and are under study. In the host-mediated assay, uninduced congenitally athymic BALB/c ("nude") mice were used. Equally reproducible results were found using BALB/c mice recently.[17] In either case, the effects of inducers for metabolic activation are yet to be determined.

[b] Agents studied are denoted either as carcinogenic or noncarcinogenic or uncertain based primarily on published data from USPHS[4] and IARC[19] regardless of "carcinogenic potency". Carcinogenicity of many compounds is not yet available. The search for such data is admittedly neither exhaustive nor updated. The data are compiled from all agents studied, excluding those whose carcinogenicity is either unknown or uncertain. Thus, only 42 out of 79 agents studied are compiled in this table.

[c] Known carcinogens are mutagenic and noncarcinogens are nonmutagenic in CHO/HGPRT assays, i.e., MNNG, ICR-292, Ni, B(a)P, hycanthone, UV etc.

[e] Known carcinogens are nonmutagenic in CHO/HGPRT assays, e.g., formaldehyde.

[f] Known noncarcinogens are mutagenic in CHO/HGPRT assays, e.g., ICR-191.

determined in diversified ionizing and nonionizing physical agents such as X-ray, UV light, sunlight, etc., as well as in various chemicals, including alkylating agents, ICR compounds, polycyclic hydrocarbons, nitrosamines, metallic compounds, etc.; a total of 79 were examined, of which 43 have documented carcinogenicity or noncarcinogenicity. Mutagenicity as determined in the CHO/HGPRT assay appears to correlate well (40/42 [95.24%]) with the reported carcinogenicity in animals within these limited chemical groups.

To insure that the determination of mutagenicity using the CHO/HGPRT system can be employed as a valid prescreen for environmental carcinogens, we have extended our studies to include various other major groups of experimental chemical carcinogens and their structurally related noncarcinogenic congeners.[20,26,33] We believe the results of this validation study should give a further assessment of the utility of the CHO/HGPRT system in both experimental mutagenesis and genetic toxicology.

REFERENCES

1. Alford, B. L. and Barnes, E. M., Hypoxanthine transport by cultured Chinese hamster lung fibroblasts, *J. Biol. Chem.*, 251, 4823, 1976.
2. Ames, B. N., McCann, J., and Yamasaki, E., Methods for detecting carcinogens and mutagens and Salmonella/mammalian-microsome mutagenicity test, *Mutat. Res.*, 31, 347, 1975.
3. Chasin, L. A., The effect of ploidy on chemical mutagenesis in cultured Chinese hamster cells, *J. Cell. Physiol.*, 82, 299, 1973.
4. Chu, E. H. Y. and Malling, H. V., Chemical induction of specific locus mutations in Chinese hamster cells in vitro, *Proc. Natl. Acad. Sci. U.S.A.*, 61, 1306, 1968.
5. Chu, E. H. Y. and Powell, S. S., Selective systems in somatic cell genetics, *Adv. Hum. Genet.*, 7, 189, 1976.
6. Couch, D. B. and Hsie, A. W., Mutagenicity and cytotoxicity of congeners of two classes of nitroso compounds in Chinese hamster ovary cells, *Mutat. Res.*, 57, 209, 1978.
7. Couch, D. B. and Hsie, A. W., Dose-response relationships of cytotoxicity and mutagenicity of monofunctional alkylating agents in Chinese hamster ovary cells, *Mutat. Res.*, 38 (Abstr.), 399, 1976.
8. Couch, D. B., Forbes, N. L., and Hsie, A. W., Comparative mutagenicity of alkylsulfate and alkanesulfonate derivatives in Chinese hamster ovary cells, *Mutat. Res.*, 57, 217, 1978.
9. Deaven, L. and Peterson, D., The chromosome of CHO, an aneuploid Chinese hamster cell line: G-band, C-band and autoradiographic analysis, *Chromosoma*, 41, 129, 1973.
10. Gabridge, M. G. and Legator, M. S., A host-mediated assay for the detection of mutagenic compounds, *Proc. Soc. Exp. Biol. Med.*, 130, 831, 1969.
11. Hsie, A. W. and Brimer, P. A., unpublished results.
12. Hsie, A. W., Brimer, P. A., Machanoff, R., and Hsie, M. H., Further evidence for the genetic origin of mutations in mammalian somatic cells: the effects of ploidy level and selection stringency on dose-dependent chemical mutagenesis to purine analogue resistance in Chinese hamster ovary cells, *Mutat. Res.*, 45, 271, 1977.
13. Hsie, A. W., Brimer, P. A., Mitchell, T. J., and Gosslee, D. G., The dose-response relationship for ethyl methanesulfonate-induced mutations at the hypoxanthine-guanine phosphoribosyl transferase locus in Chinese hamster ovary cells, *Somat. Cell Genet.*, 1, 247, 1975.
14. Hsie, A. W., Brimer, P. A., Mitchell, T. J., and Gosslee, D. G., The dose-response relationship for ultraviolet light-induced mutations at the hypoxanthine-guanine phosphoribosyl transferase locus in Chinese hamster ovary cells, *Somat. Cell Genet.*, 1, 383, 1975.
15. Hsie, A. W., Couch, D. B., and Brimer, P. A., unpublished results.
16. Hsie, A. W., Li, A. P., and Machanoff, R., A fluence-response study of lethality and mutagenicity of white, black, and blue fluorescent lights, sunlamp, and sunlight irradiation in Chinese hamster ovary cells, *Mutat. Res.*, 45, 333, 1977.
17. Hsie, A. W., Machanoff, R., Couch, D. B., and Holland, J. M., Mutagenicity of dimethylnitrosamine as studied by a quantitative host-mediated specific-locus mutation assay in Chinese hamster ovary cells, *Mutat. Res.*, 51, 77, 1978.
18. Hsie, A. W., Machanoff, R., Couch, D. B., and Holland, J. M., A quantitative host-mediated mutational assay of the hypoxanthine-guanine phosphoribosyl transferase locus in Chinese hamster ovary cells, *Mutat. Res.*, (Abstr), Annual Meeting of the Environmental Mutagen Society, Colorado Springs, Colorado, 1977, in press.

19. *IARC Monograph on the Evaluation of Carcinogenic Risk of Chemicals to Man,* Vol 1 to 10, International Agency for Research on Cancer, Lyons, France, 1972 to 1976.
20. **Ishidate, M., Jr. and Odashima, S.,** Chromosome tests with 134 compounds on Chinese hamster cells in vitro: a screening for chemical carcinogens, *Mutat. Res.,* 48, 337, 1977.
21. **Kao, F. T. and Puck, T. T.,** Induction and isolation of nutritional mutants in Chinese hamster cells, *Proc. Natl. Acad. Sci. U.S.A.,* 60, 1275, 1968.
22. **Kao, F. T. and Puck, T. T.,** Quantitation of mutagenesis by physical and chemical agents, *J. Cell. Physiol.,* 74, 245, 1969.
23. **Krahn, D. F. and Heidelberger, C.,** Liver homogenate-mediated mutagenesis in Chinese hamster V79 cells by polycyclic aromatic hydrocarbons and aflatoxins, *Mutat. Res.,* 46, 27, 1977.
24. **Lyon, M. F.,** Sex chromatin and gene action in the mammalian X chromosome, *Am. J. Hum. Genet.,* 14, 135, 1962.
25. **McCann, J. and Ames, B. N.,** Detection of carcinogens as mutagens in the Salmonella/microsome test: assay of 300 chemicals: discussions, *Proc. Natl. Acad. Sci. U.S.A.,* 73, 950, 1976.
26. **McCann, J., Choi, E., Yamasaki, E., and Ames, B. N.,** Detection of carcinogens as mutagens in the Salmonella/microsome test: assay of 300 chemicals, *Proc. Natl. Acad. Sci. U.S.A.,* 72, 5135, 1975.
27. **Nettesheim, P.,** unpublished results.
28. **O'Neill, J. P., Brimer, P. A., Machanoff, R., Hirsch, G. P., and Hsie, A. W.,** A quantitative assay of mutation induction at the hypoxanthine-guanine phosphoribosyl transferase locus in Chinese hamster ovary cells: development and definition of the system, *Mutat. Res.,* 45, 91, 1977.
29. **O'Neill, J. P., Couch, D. B., Machanoff, R., San Sebastian, J. R., Brimer, P. A., and Hsie, A. W.,** A quantitative assay of mutation induction at the hypoxanthine-guanine phosphoribosyl transferase locus in Chinese hamster ovary cells (CHO/HGPRT system): utilization with a variety of mutagenic agents, *Mutat. Res.,* 45, 103, 1977.
30. **O'Neill, J. P., Fuscoe, J. N., and Hsie, A. W.,** Mutagenicity of heterocyclic nitrogen mustards (ICR compounds) in cultured mammalian cells, *Cancer Res.,* 38, 506, 1978.
31. **O'Neill, J. P. and Hsie, A. W.,** Chemical mutagenesis of mammalian cells can be quantified, *Nature (London),* 269, 815, 1977.
32. **Peck, R. M., Tan, T. K., and Peck, E. B.,** Pulmonary carcinogenesis by derivatives of polynuclear aromatic alkylating agents, *Cancer Res.,* 36, 2423, 1976.
33. **Pienta, R. J., Poiley, J. A., and Lebherz, W. B.,** Morphological transformation of early passage golden Syrian hamster embryo cells derived from cryopreserved primary cultures as a reliable in vitro bio-assay for identifying diverse carcinogens, *Int. J. Cancer,* 19, 642, 1977.
34. **Puck, T. T.,** *The Mammalian Cell as a Microorganism,* Holden-Day, San Francisco, 1972.
35. **Puck, T. T. and Kao, F. T.,** Treatment of 5-bromodeoxyuridine and visible light for isolation of nutritionally deficient mutants, *Proc. Natl. Acad. Sci. U.S.A.,* 58, 1227, 1967.
36. **Roy-Burman, P.,** Analogues of nucleic acid components, *Recent Results Cancer Res.,* 25, 9, 1970.
37. **Russell, L. B.,** Genetics of mammalian sex chromosomes, *Science,* 133, 1795, 1961.
38. **San Sebastian, J. R., Couch, D. B., and Hsie, A. W.,** unpublished results.
39. **Sharp, J. D., Capeechi, N. E., and Capeechi, M. R.,** Altered enzymes in drug resistant variants of mammalian tissue culture cells, *Proc. Natl. Acad. Sci. U.S.A.,* 70, 3145, 1973.
40. **Siminovitch, L.,** On the nature of heritable variation in cultured somatic cells, *Cell,* 7, 1, 1976.
41. **United States Public Health Service,** Survey of Compounds which Have Been Tested for Carcinogenic Activity, Publ. No. 149, *U.S. Public Health Service,* Washington, D.C., 1972—1973.

THE USE OF CULTURED MAMMALIAN CELL TRANSFORMATION SYSTEMS TO IDENTIFY POTENTIAL CARCINOGENS

D. F. Krahn

TABLE OF CONTENTS

I.	Introduction	55
II.	Hamster Embryo Cells	56
III.	Hamster Embryo Cell Transplacental Assay	57
IV.	Hamster Cell Line	58
	A. BHK_{21}	58
V.	Mouse Cell Lines	59
	A. BALB/3T3	59
	B. C3H/10T1/2	59
	C. M2	60
VI.	Rat Cell Line	61
VII.	Systems Combining Virus and Chemicals	61
	A. Hamster Embryo Cells	61
	B. Rat Embryo Cell Line	61
	C. Mouse Embryo Cell Line	62
VIII.	Human Cell Systems	63
IX.	Epithelial Cell Systems	63
X.	Conclusions	63
References		64

I. INTRODUCTION

There is strong evidence that chemicals play an important role in the causation of the majority of human cancers.[24,73,74] The identification of those chemicals that are carcinogens, followed by action to minimize the exposure of humans to them, should be an effective means to reduce the cancer incidence.

We are now confronted with the need for methods that provide a rapid and reliable indication of the potential carcinogenic hazard that a chemical may represent for humans. The conventional lifetime carcinogenicity studies in animals require sizeable facilities to house the animals for testing and are extremely expensive, costing between $100,000 to $500,000 per chemical. Furthermore, by the time the pathology is completed, it may be 3 years before the data from a normal rat study are available. In addition to the space, cost, and time concerns, the lifetime studies in rodents or other mammals are not 100% reliable for predicting the potential hazard for humans. Thus, in order to evaluate the large number of chemicals that require testing, it is necessary to develop and utilize rapid and inexpensive bioassays that identify potential carcinogens. The cultured mammalian cell transformation systems make up one class of the so-called short-term tests that show promise as predictive bioassays. In addition to their potential for use as prescreens, these systems continue to be valuable tools for the analysis of the mechanism(s) of chemical carcinogenesis.

The use of cultured mammalian cells to study chemical carcinogenesis has been reviewed by Castro and DiPaolo,[13] Heidelberger,[33,34] and Weinstein et al.[72] In this review my intent is to discuss the cell transformation systems that are presently considered to be the most promising candidates for use as routine bioassays to identify potentially carcinogenic chemicals.[25] The emphasis is on the nature of the individual systems and the extent to which they have been evaluated with diverse chemical carcinogens.

II. HAMSTER EMBRYO CELLS

The use of cells derived from hamster embryo tissue to study chemically induced transformation has been thoroughly reviewed by DiPaolo[22] and Heidelberger.[33,34] Primary hamster embryo cell cultures are prepared by treating minced embryo tissue with trypsin and plating the cells thus obtained. The primary and secondary cultures are considered to be composed predominantly of diploid fibroblasts. These cells have a limited lifespan and feeder cells must be present in order to obtain reasonable plating efficiencies when low numbers of cells are plated. The feeder cells are generally primary or secondary cultures of hamster or rat embryo cells that have been X-irradiated. The number of cells plated per dish in a transformation experiment is low enough to permit individual colony formation. The colonies are stained 7 to 10 days after treatment and then examined for the transformed morphology. Cells in normal colonies pile up to a very limited extent, whereas transformed colonies exhibit extensive piling up. These transformed colonies are composed of small fusiform-shaped cells that appear in a random crisscross pattern that is especially evident at the periphery of the colony.

The first clear demonstration of chemical carcinogen-induced transformation was reported by Berwald and Sachs[10] with hamster embryo cells. These workers noted that mass cultures treated with 3-methylcholanthrene (3-MC) and benzo[a]pyrene (B[a]P) and then plated on a layer of irradiated rat embryo feeder cells gave rise to colonies that exhibited a crisscross, piled-up morphology. No transformed colonies appeared when the cultures were treated with urethane or solvent. Huberman and Sachs[35] presented data with B[a]P that indicated that the increased frequency of transformed colonies is a result of induction and not merely due to the selection of colonies resistant to toxicity.

Numerous carcinogenic chemicals have been shown to induce morphological transformation in the hamster embryo system. Huberman et al.[38] reported that dimethylnitrosamine-treated secondary cultures acquired an indefinite life span and a shortened doubling time, but did not exhibit the characteristic crisscross morphology or become tumorigenic until several months after treatment. Kuroki and Sato[47] observed that transformation occurred when hamster embryo cell primary or secondary cultures were treated with 4-nitroquinoline-1-oxide (4NQO), or two carcinogenic derivatives, and then periodically passaged. No transformation was observed when the cultures were treated with two noncarcinogenic derivatives of 2NQO. DiPaolo et al.[17] demonstrated that the frequencies of transformed colonies which were produced after treatment with a series of polycyclic hydrocarbons closely paralleled their carcinogenic activities in vivo and that the cells of the transformed colonies were tumorigenic in hamsters.[18] Several epoxides of polycyclic hydrocarbons have been found to produce greater frequencies of transformed colonies than the parent compounds or other derivatives.[31,37,50] DiPaolo et al.[19] demonstrated that the carcinogens aflatoxin B_1, N-acetoxy-2-acetylaminofluorene (N-AcAAF), N-methyl-N'-nitro-N-nitrosoguanidine (MNNG), and methylazoxymethanol produced morphological transformation. Slight activity was observed with 2-acetylaminofluorene, but the carcinogens urethane and diethylnitrosamine (DEN) were not active in this system. Huberman et al.[38] noted that transformation was induced by N-AcAAF and N-hydroxy-2-acetylaminofluorene, but not by the parent compound, 2-acetylaminofluorene (AAF).

Recently, Pienta et al.[59] published results of a study in which an attempt was made to identify factors that affect the reproducibility of the hamster embryo system. In addition, the response of the system to a large number of diverse chemicals was evaluated. They observed that not all separately prepared batches of embryo cells could be transformed by 3-MC. When frozen aliquots of 3-MC-sensitive batches were thawed and retested, 3-MC again induced transformation, although the 3-MC-induced transformation frequencies varied among different aliquots of a given batch. In light of these findings, Pienta and co-workers decided to perform experiments with cells from frozen aliquots of pretested target and feeder cells. There were 87 chemicals tested that included direct alkylating agents, polycyclic hydrocarbons, nitrosamines and amides, aromatic amines and aminoazo dyes, metals, and miscellaneous compounds. The ability to induce transformation correlated with their reported in vivo carcinogenic activities for about 91% of the chemicals tested. No false positive results were observed and no transformed colonies appeared in control cultures. A chemical was classified as being active in the system if one or more transformed colonies were observed. Carcinogens that require metabolism for activity such as AAF, DEN, 3-methoxy-4-aminoazobenzene, 2-nitrofluorene, urethane, auramine, natulan, 4-aminoazobenzene, and p-rosaniline did not induce transformation. However, AAF did induce transformation when a rat liver homogenate-activation system similar to that described by Ames et al.[2] was used. Preliminary experiments indicated that all members of the above list except 4-aminoazobenzene induce transformation when the treatment is performed in the presence of an activation system.[75] All of the transformed colonies tested were tumorigenic.

In studies directed toward understanding the relationship between transformation and mutation, Huberman et al.[39] looked at the simultaneous induction of transformation and mutation in primary hamster embryo cells. They observed that the frequencies of transformation induced by B[a]P and a diol derivative were about 20 times greater than the frequencies of ouabain-resistant mutants, when calculated on the basis of cell survival. Ts'o[70] found that mutation frequencies measured by resistance of 8-azaguanine, 6-thioguanine, and ouabain were 10^2- to 10^3-fold less than the transformation frequencies, in a study using hamster embryo cells.

The Syrian hamster embryo colony assay has been modified by Casto et al.[15] so that transformation is indicated as foci upon a background of normal cells. Transformed foci appeared in cultures treated with the carcinogens AAF, aflatoxin B_1, B[a]P, β-propiolactone, dibenz[a,h]anthracene (DB[a,h]A), ethyl methanesulfonate (EMS), methylmethane sulfonate (MMS), 3-MC, and MNNG. Foci are scored in dishes stained 15 to 21 days after treatment. These authors suggest that the focus assay has the advantage of reducing the number of areas that have to be examined relative to the colony assay and is not subject to the relatively low plating efficiencies that occur with the colony assay.

III. HAMSTER EMBRYO CELL TRANSPLACENTAL ASSAY

A hamster embryo cell transformation system in which transformed colonies are observed in embryo cell cultures that are obtained from the embryos of treated pregnant hamsters has been developed by DiPaolo et al.[19,21] Morphological transformation was observed in either mass cultures or in sparsely seeded cultures in the presence of feeder cells. No transformation was observed in cultures obtained from pregnant animals that were treated with solvent or noncarcinogens, but each of the 12 carcinogens tested caused the appearance of transformed colonies. This may prove to be a useful bioassay for detecting chemicals that require metabolic activation and has the potential for studying the interaction of multiple chemicals that may be enzyme inducers or cocarcinogens. Several factors, which include the metabolic capability of the embryonic cells, the stability of metabolites generated by the mother, and the ability of the chemical or its metabolites to reach the embryo, influence whether a chemical will be active in this assay.

IV. HAMSTER CELL LINE

A. BHK_{21}

The BHK_{21} cell line was derived from baby Syrian hamster kidney tissue and has been used as a tool for the study of transformation induced by virus and chemicals. The BHK_{21} cells have been reported to possess a diploid modal chromosome number.[68] In this system, transformed cells are generally identified by their ability to form colonies in soft agar. Treated cells are immediately plated into soft agar and visible colonies counted after a 2- to 4-week incubation period. The ability of these transformed cells to form colonies when placed in agar so soon after treatment is in sharp contrast to what is observed with most other cell transformation systems. In the majority of other systems, transformed cells, as identified by morphological criteria, do not acquire the ability to grow in soft agar nearly as rapidly.

Chemically induced transformation was first demonstrated with the BHK_{21} cl 13 line by DiMayorca et al.[16] and Mishra and DiMayorca.[55] These workers found that treatment with DMN or nitrosomethylurea gave transformation frequencies of about 6×10^{-6} and that very few spontaneous transformants appeared. Cells from each of the transformed colonies tested were tumorigenic when inoculated into hamsters. The parent, untreated cells were also tumorigenic. However, the number of cells required per inoculum and the length of the period before tumor formation were greater for untreated cells than for transformed cells. These workers also noted that the transformed phenotype was dependent on temperature; the transformed phenotype persisted at 38.5°C, but was lost when the temperature was shifted to 32°C.

Recently Purchase et al.[61] evaluated the response of the BHK_{21} system, as well as several other short-term assays, to 58 carcinogens and 62 noncarcinogens, as defined by data from whole-animal carcinogenicity studies. Their list included polycyclics, arylamines, alkylating agents, and miscellaneous chemicals. Each compound was tested in the presence of a rat liver homogenate-activation system similar to that described by Ames et al.[3] They observed that 91% of the carcinogens caused transformation and that 97% of the noncarcinogens were inactive. The details of the test protocol and a complete list of test chemicals have not yet been published. Carcinogens that did not induce transformation include 2-naphthylamine, vinyl chloride, and diethylstilbesterol.

Quantitative studies of chemically induced transformation in the BHK_{21} cell system have been reported by Ishii et al.[40] Experiments were performed with cells of a subclone of BHK_{21} cl 13 that are capable of dividing several times in soft agar and exhibit a spontaneous transformation frequency of 5 to 10×10^{-6}. Concentration-dependent increases in cytotoxicity and transformation frequency were observed with 4-NQO, N-AcAAF, MNNG, and ultraviolet light. Increases in the transformation frequency of 50- to 100-fold were observed in cultures treated with the three chemicals. The carcinogen, AAF, was not active, presumably because the cells lack the ability to convert it to a reactive metabolite. About one half of the transformed clones that were tested gave tumors in hamsters; none of the nontransformed cells were tumorigenic under the conditions used.

On the basis of experiments performed with nitrosomethylurea and 4-NQO, Bouck and DiMayorca[11] have concluded that mutation probably is responsible for the transformation of BHK_{21} cells. The reasons for their conclusion are that the spontaneous and chemically induced transformation frequencies are in the range observed for mutation frequencies in many systems, temperature-sensitive transformants have been observed, the transformed phenotype is stably expressed, and the reversion frequency is low.

The BHK_{21} system has the advantages that cells can be plated in soft agar immediately after treatment and that transformation is readily indicated by the number of colonies that appear. Since an expression time in mass culture is not required prior to plating in agar, the transformation frequencies can be confidently expressed on a cell survival basis. A possible disadvantage of this system is associated with the fact that the parent population is tumorigenic if enough cells are inoculated into hamsters.[16,67,68] Since the parent cells appear to possess some characteristics of the transformed state, Ishii et al.[40] have concluded that the induction of transformation as defined in this system may represent only one or a few of the steps involved in carcinogenesis. It is generally accepted that several steps are involved in carcinogenesis in vivo[26] and it seems likely that transformation in vitro is also a multistep process. If transformation in the BHK_{21} system does not represent the entire transformation process, this system may not be sensitive to certain potential carcinogens.

V. MOUSE CELL LINES

A. BALB/3T3

DiPaolo et al.[20] have been able to demonstrate chemically induced transformation with subclones of a BALB/3T3 cell line.[1] These cells exhibit contact inhibition of cell division, a flat epithelial-like morphology, and are aneuploid (subtetraploid). Transformation, as indicated by a piling-up of fibroblast-like cells that exhibit a crisscross morphology, can be scored by examining discrete colonies 10 to 11 days after treatment or foci upon a cell monolayer at 3 weeks. Transformation was induced by 7,12-dimethylbenz[a] anthracene (DMBA), 3-MC, MNNG, aflatoxin B_1, and N-AcAAF, but not by DEN or the noncarcinogens, anthracene and pyrene. No transformed colonies were observed in control cultures. The trasformation frequencies induced by certain chemicals represented several percent of the initial cell inoculum and were observed to increase with increasing concentrations of carcinogen. Transformed cells, but not cells with normal morphology, produced tumors when inoculated into BALB/c mice.

Kakunaga[43] has independently transformed cells of a subclone of the BALB/3T3 line[1] with chemical carcinogens. The clone used in this study showed a high degree of contact inhibition and a low ($<5 \times 10^{-6}$ transformants per cell) incidence of spontaneous transformation. Treatment with solvent, or with the noncarcinogenic compounds 4-nitroquinoline, 4-aminoquinoline-1-oxide, phenanthrene, and pyrene, did not result in the appearance of any transformed foci. However, the carcinogens 4-NQO, 3-MC, B[a]P, and MNNG all induced the appearance of foci of piled-up cells which exhibited crisscross morphology. Cultures were stained and examined 4 weeks after treatment. The transformation frequencies increased with concentration within a certain range. Cell lines derived from the transformed foci maintained the transformed morphology and exhibited an increased saturation density, the ability to grow in soft agar, and the ability to produce tumors upon inoculation into newborn or adult mice. The increased transformation frequencies that were observed in the treated cultures were not due to a selection of preexisting transformed cells. More recently, Kakunaga[44] has demonstrated the requirement for cell division for both the fixation and expression of the transformed state.

To date, published work with these cells has dealt with only a few chemicals. Members of many different classes of chemicals must be evaluated before the utility of this assay as a general screen will be known. The fact that the carcinogen DEN is not transforming[20] suggests that this system probably requires a supplementary activation system.

B. C3H/10T1/2

A transformable cell line derived from C3H mouse embryo tissue, designated C3H/10T1/2, was developed in Heidelberger's laboratory at the University of Wisconsin.[63] These cells are aneuploid (subtetraploid) and exhibit contact inhibition of cell division. Cells growing in plastic dishes are treated and then incubated for 6 weeks to assess the transforming activity of a chemical. During this period the growth medium is periodically changed. At the end of 6 weeks the dishes are stained and examined for transformed foci. Transformation is indicated by piling-up and the crisscrossing of cells at the periphery of the focus. Cells cloned from transformed foci, but not nontransformed cells, are tumorigenic in syngeneic C3H mice. The incidence of spontaneous transformation is very low.

A variety of chemicals have been found to induce transformation in this system. Concentration-dependent increases of the transformation frequency were observed with the polycyclic hydrocarbon carcinogens, 3-MC and DB[a,h]A,[62] DMBA[9,62] and B[a]P.[32] Bertram and Heidelberger[8] reported the induction of transformation in a synchronous population of C3H/10T1/2 cells by very cytotoxic concentrations of MNNG. They found that the transforming activity of this chemical exhibited a marked cell-cycle dependency with a sensitive phase for malignant transformation during the 4 hr prior to the onset of S phase. Benedict et al.[5] observed that a crude cigarette-smoke condensate, a reconstituted sample, and two specific fractions induced transformation. It is interesting to note that these two active fractions were found to be mutagenic for *Salmonella*[46] but that two different fractions that displayed mutagenic activity did not transform. Benedict[6] found that two hair-dye components, 2-nitro-*p*-phenylenediamine and 4-nitro-*o*-phenylenediamine, induced transformation in C3H/10T1/2 cells. The cancer chemotherapeutic drugs, 5-fluorodeoxyuridine[41] and 1-β-D-arabinofuranosylcytosine,[42] induced transformation when treatment occurred during the S phase of a synchronous culture. In another study Benedict et al.[7] studied the transforming activity of ten chemotherapeutic drugs that included alkylating agents, antimetabolites, antibiotics, and natural products. In this study, the chemicals that are known to be carcinogenic in vivo produced transformation, whereas those chemicals that are thought to be noncarcinogenic were inactive.

The C3H/10T1/2 system also appears to be sensitive to the action of certain promoters. Mondal et al.[56] demonstrated that the initiation and promotion stages of carcinogenesis that occur in the C3H/10T1/2 cells are similar to those observed in mouse skin. A significant enhancement of transformation occurred when cells treated with a concentration of 3-MC or B[*a*]P, which did not transform by itself, were treated 4 days later with the in vivo promoters tetradecanoyl phorbol acetate (TPA) and phorbol didecanoate, or with 4-α-phorbol didecanoate, a chemical for which the in vivo promoting activity is not clear. These phorbol derivatives did not cause transformation by themselves. Phorbol, which is not a promoter in mouse skin, was inactive in this system. The promoter, TPA, did not exhance the transforming activity of DMBA, possibly because the concentration of DMBA used was transforming by itself. The noncarcinogenic polycyclic hydrocarbons, anthracene and phenathrene, did not transform C3H/10T1/2 cells, either alone or with TPA treatment. This demonstration of promoter activity opens up the possibility that cell culture systems, such as the C3H/10T1/2 system, might be useful for identifying promoters in the environment. However, considerable research is necessary before a confident assessment of the usefulness of such systems for this purpose can be made.

Promotion by TPA and croton oil has also been studied by Lasne et al.[48] in a rat embryo cell system. They observed that the morphological transformation induced by DMBA, B[*a*]P, and benzo[*e*]pyrene was accelerated when the polycyclic hydrocarbon treatment was followed by a promoter. Unlike the observations of Mondal et al.[56] TPA alone induced transformation in this study. Mouse embryo C3H/10T1/2 cells have also been transformed by ultraviolet light followed by treatment with a phorbol ester.[57]

There are factors that affect the quantitation of the C3H/10T1/2 cell system that require further investigation. The frequency of transformed foci that arise has been shown to be dependent on cell density.[32,62,69] Betram[9] has also observed that the density of nontransformed cells influences the expression of transformed cells and that this influence can be modulated by changes of serum concentration. In order to obtain more meaningful concentration-response data and perhaps to increase the sensitivity of the system, this cell density-related phenomenon should be further investigated.

Consistent with their susceptibility to polycyclic hydrocarbon-induced cytotoxicity and transformation, the C3H/10T1/2 cells do possess aryl hydrocarbon hydroxylase activity.[58] However, it remains to be demonstrated whether this system will respond to nitrosamines, aromatic amines, azo dyes, and other classes of chemicals that require metabolic activation. It is likely that these cells will be unable to activate members of at least some of these classes, and, therefore, the capability for activation will have to be provided in order to make this assay useful as a general screen.

C. M2

Marquardt[51] has developed a line of mouse fibroblasts that are derived from C3H mouse prostate tissue. These cells, designated M2, have been transformed by several polycyclic hydrocarbons and derivatives,[51,52,54] MNNG, and the carcinogenic antitumor antibiotics adriamycin and daunomycin (Marquardt et al.[53]). No reports concerning the sensitivity of this system to other types of chemicals have yet appeared.

VI. RAT EMBRYO CELL LINE

The transforming activity of polycyclic hydrocarbons was investigated by Rhim and Heubner[64] in a Fischer rat embryo cell line. Cell cultures were subjected to treatment with the test chemical for 7 days and subdivided 7 days later. This subdivision was repeated every 7 to 10 days thereafter. Benzo[a]pyrene induced observable transformation, as indicated by the appearance of crisscrossing, randomly oriented, spindle-shaped cells, 40 to 43 days after treatment. Transformed clones, but not clones with normal morphology, were tumorigenic in newborn rats. Similar results were obtained with DMBA and 3-MC.

VII. SYSTEMS COMBINING VIRUS AND CHEMICALS

A. Hamster Embryo Cells

The interaction of viruses and chemicals in carcinogenesis has been reviewed by Casto and DiPaolo[13] and DiPaolo and Casto.[23] There are several cultured mammalian cell transformation systems in which transformation that is observed following treatment with a chemical carcinogen is dependent on the presence of virus.

The transformation of hamster embryo cells by simian adenovirus, SA7, is quantitatively enhanced when the inoculation of virus is preceded by treatment with certain chemical carcinogens. These studies have been reviewed by Casto et al.[14] and DiPaolo and Casto.[23] In these studies, primary hamster embryo cells were treated with the test chemical, and then infected with SA7 and replated. The frequency of SA7 transformed foci was determined 21 to 25 days later. Transformed foci were not seen in dishes that did not receive virus or in dishes treated with chemical alone. The enhancement of viral transformation is determined by dividing the frequency of transformation in a cell population that receives chemical and SA7 virus by the frequency that is observed in a population that receives SA7 alone. The degree of enhancement that is observed is dependent on the time of chemical treatment in relation to the time of virus addition. Alfatoxin B_1 and the carcinogenic hydrocarbons B[a]P, 3-MC, DB[a,h]A, DB[a,c]A, and DMBA enhanced SA7 transformation when added prior to virus, but inhibited transformation when added after virus adsorption and replating. The weak carcinogen, benzo[e]pyrene, gave a slight enhancement, but the noncarcinogens, pyrene, phenanthrene, and perylene had no effect. The enhancement of SA7 transformation by cytosine arabinoside, caffeine, and 6-acetoxy-benzo[a]pyrene was greater when the chemical treatment was made after the addition of virus. An enhancement of transformation was induced when treatment was before or after virus addition with the chemicals β-propiolactone, MMS, N-AcAAF, MNNG, and methylazoxymethanol. The carcinogens AAF, DEN, and DMN did not enhance SA7 transformation, presumably because the hamster embryo cells are unable to activate these compounds. In the studies discussed by Casto et al.[14] the ability of the test chemicals to induce DNA repair-synthesis and changes in DNA sedimentation was also evaluated. Casto[12] has observed that certain DNA base analogues also enhance SA7 transformation. With a similar system, Ledinko and Evans[49] found that MNNG, caffeine, and hydroxylamine enhance adenovirus 12 transformation.

It has been proposed[13,14] that chemical carcinogens enhance viral-induced transformation by providing additional sites in DNA at which viral genetic material can be integrated. In order for a chemical to be active in a virus transformation the enhancement assay, it appears that the chemical must induce the formation of breaks or gaps in DNA and that the virus be present while the DNA is damaged.

B. Rat Embryo Cell Lines

Numerous chemicals have been found to transform cultured cells derived from rat embryo tissue that have been infected with RNA leukemia virus.[28,29,60] In the procedure utilized by Freeman et al.[28,29] a mass culture of rat embryo cells is treated with the test chemical 1 week after being infected with rat leukemia virus. The chemical is allowed to remain in the culture for 1 week. The treated culture is split into a vertical culture and a holding culture. The holding culture is incubated and examined for the appearance of transformed foci. The vertical culture is split every 2 weeks to form a vertical culture and a holding culture. Up to three subdivisions of this sort may be carried out. Both the vertical and holding series cultures are examined for the presence of transformed foci. The point at which transformation is observed depends on the activity of the test chemical. The assay may continue for up to 8 weeks. Transformed foci are characterized by the piling-up of randomly oriented, crisscrossed cells.

Freeman et al.[28,29] have demonstrated that the susceptibility of the rat embryo cells to chemically induced transformation varies with the age of the culture. For 35 to 60 population doublings the cultures are diploid, exhibit contact inhibition, are not tumorigenic, and can be transformed by a combination of rat leukemia virus and chemical carcinogen, but not by either agent alone. During the next 50 to 60 generations the cells exhibit the same properties as the low passage cells, except that they are heteroploid and can be transformed by chemical carcinogens alone. However, the transformed cells were shown to possess the gs-1 antigen of endogenous rat leukemia virus. As the age of the culture increases, the spontaneous transformation frequency gradually increases.

The transformation assay protocol was designed to include periodic subdivisions in order to provide a selective advantage to the noncontact-inhibited transformed cells. This selective advantage serves to magnify the transforming activity of a chemical and, thus serves to increase the sensitivity of the assay. Since the treated cells are periodically replated throughout an experiment, the actual number of transformed foci that a chemical induces is of little importance for comparing the potencies of different chemicals. Such comparisons are valid only if foci that appear in the initial holding culture are considered. In this assay quantitative information is indicated by comparing the optimal transforming concentrations.

In a single study, Freeman et al.[28,29] have examined the response of rat embryo cells in the heteroploid phase to 35 chemicals that include polycyclic hydrocarbons, azo dyes, aromatic amines, and miscellaneous chemicals. In general, transformation was induced by the known carcinogens, but not by chemicals known to be noncarcinogenic. A few chemicals gave variable results among different experiments. These cases of variation were explained by the failure to utilize an optimal transforming-concentration (4-aminobiphenyl and propane sultone) or by the apparent weakness of the carcinogen tested (acetamide). The aromatic amine, AAF, was consistently negative, presumably because the cells were unable to metabolize it to its active form.

In another study, both rat and mouse embryo cell lines were transformed by benzene extracts of airborne particulate matter.[35] Price et al.[60] have observed that 3-MC would transform rat embryo cell cultures if the treatment occurred after virus infection, but not if the 3-MC treatment took place before virus infection.

Freeman et al.[29] consider carcinogenic chemicals and C-type RNA viruses to act as cocarcinogenic agents in the rat embryo cell system. There is as yet no proof that the expression of endogenous C-type virus is a necessary prerequisite or an absolute result of transformation. There is a definite association between C-type virus expression and transformation, but it is possible that these events are independent of one another.

Recently, Waters et al.[71] obtained evidence in studies with 4-nitroquinoline-1-oxide that the capability for postreplication repair is diminished in rat embryo cells infected with leukemia virus compared with uninfected cells. It is suggested that leukemia virus sensitizes the rat cells to carcinogens by affecting the postreplication repair of DNA damaged by the carcinogenic chemical.

C. Mouse Embryo Cell Line

A line of NIH Swiss mouse embryo cells infected with a nontransforming-AKR leukemia virus can also be transformed by chemical carcinogens.[65] The transformation assay protocol is similar to the one employed in the rat embryo cell system described above. Rhim et al.[65] have evaluated the response of almost 30 compounds that represent several different types of chemical carcinogens and noncarcinogenic analogues. Foci of transformed cells consisted largely of crisscrossing randomly orientated, and piled-up spindle-shaped cells. In the majority of cases, the in vivo carcinogenic activity of a chemical was reflected in its ability to transform the AKR-infected mouse embryo cells.

However, transformation was not induced by the carcinogens DB[*a,h*]A, 3-methyl-4-dimethylaminoazobenzene, and H-hydroxy-2-acetylaminofluorene, whereas the reported noncarcinogen 2-biphenylamine did transform. The noncarcinogen phenanthrene gave variable results.

VIII. HUMAN CELL SYSTEMS

There are to the author's knowledge only a few instances in which the transformation of human cells in culture by chemicals has been reported. Benedict et al.[4] studied two clones obtained from morphologically transformed foci in urethane-treated cultures of cells obtained from a patient with von Recklinghausen's disease, a familial disorder characterized by multiple fibromas with a high predisposition to malignant transformation in vivo. One of the clones was derived from a urethane-treated population that was preinfected with feline leukemia virus. Aside from the morphological characteristics, the transformed clones grew in soft agar, exhibited an increased level of fibrinolytic activity, and were tumorigenic when inoculated into newborn hamsters treated with antithymocyte serum. The control cells exhibited none of these properties.

Rhim et al.[66] have reported the transformation of a human osteosarcoma cell line by MNNG. This line of cells exhibits several characteristics of transformed cells, but does not produce tumors in nude athymic mice. Foci of crisscrossed, random-oriented cells appeared in the MNNG-treated cultures, and cells from these cultures were tumorigenic in nude athymic mice.

Recently, Freedman and Shin[27] were able to detect colony formation in methylcellulose after diploid human embryo cells or primary human fibroblasts were treated with MNNG. The spontaneous frequency of colony formation was less than 10^{-8}, while in treated populations the frequency was about 10^{-5} per mutagenized cell. The clones that grew in methylcellulose also formed foci of multilayered crisscross cells, but did not possess the ability to divide indefinitely and did not produce tumors in nude athymic mice.

At this time, there is no human cell transformation system that appears ready for use as a routine prescreen.

IX. EPITHELIAL CELL SYSTEMS

Since the majority of human cancers arise from epithelial rather than fibroblastic tissues, considerable effort is being directed toward the development of transformation systems that utilize cultured epithelial cells. Much of the work in this area has been reviewed by Weinstein et al.,[72] Heidelberger,[34] and Katsuda et al.[45] The majority of work has been performed with epithelial cells obtained from rat liver, rat salivary gland, or mouse skin. There have been only a few reports on the successful transformation of cultured epithelial cells. Weinstein et al.[72] indicate that most of the present systems have the following limitations:

1. There is generally a long lag time (8 to 36 weeks) between the time of treatment and the demonstration of transformation.
2. Repeated or continuous exposure to the test chemical is often required.
3. Transformated cells do not acquire a distinctive morphology.

It is apparent that further development is required before epithelial cell systems will be ready for use as routine bioassays.

X. CONCLUSIONS

Cultured mammalian cell transformation assays continue to show promise as valuable tools for identifying suspect chemical carcinogens. Certainly much additional work is necessary to further evaluate and refine these assays. None of the present assays has reached the state of development at which neither false negative or false positive results are found. In order to have confidence in the reliability of these assays, their response to carcinogenic and noncarcinogenic members of diverse chemical classes must be known and must be reproducible among different laboratories.

Aside from the need to assess the response of these assays to diverse chemicals, there are several areas that require special attention. It is clear that the cells of nearly all of the transformation systems lack the ability to metabolize at least certain carcinogenic chemicals to their active intermediates. Therefore, the capability for metabolic activation must be provided in order to make these cell systems useful as general prescreens. It remains to be demonstrated whether the use of mammalian tissue homogenates, feeder cells, or some host-mediated approach will provide the most relevant and workable means to supply the capability for metabolic activation. Another area that requires special attention is the standardization of the systems so that data can be readily reproduced in different laboratories. Periodic variations in the frequencies of spontaneous transformants is another bothersome problem. There is also a need to develop good biochemical markers that can be used to rapidly and unequivocally identify transformed cells. Markers of this sort would reduce the need to identify transformed cells on the basis of what is, in many cases, a subjective analysis of morphology. Developments in these areas will greatly increase the utility of transformation systems for routine testing.

With certain cell transformation systems, data have been presented to indicate that the in vivo potencies of carcinogens are reflected in the degree of transformation observed in vitro. In the majority of these cases a series of chemicals belonging to a single class, such as polycyclic hydrocarbons, was evaluated. It remains to be seen how well these cell culture systems predict in vivo potency when chemicals of diverse structure, including ones that require a supplemental activation system, are evaluated.

The present cultured mammalian cell transformation assays are not considered to provide an adequate basis for making a definitive judgment on the carcinogenicity of a chemical in humans. Rather, a positive response in an assay of this type indicates that the chemical is potentially hazardous and that testing in more sophisticated bioassays is necessary. Thus, well-characterized transformation assays appear to be valuable aids in setting priorities for animal tests. In addition, these assays are potentially useful for identifying suspect carcinogens early in the development of new product candidates and in providing guidance for designing out carcinogenic activity from chemicals that are beneficial. As the mechanism(s) involved in chemical carcinogenesis become better understood, it is likely that cell transformation assays will be developed and refined so that they will play an increasingly important role in identifying potentially carcinogenic substances.

REFERENCES

1. Aaronson, S. A. and Todaro, G. J., *J. Cell Physiol.*, 72, 141, 1968.
2. Ames, B. N., Durston, W. E., Yamasaki, E., and Lee, F. D., *Proc. Natl. Acad. Sci. U. S. A.*, 70, 2281, 1973.
3. Ames, B. N., McCann, J., and Yamasaki, E., *Mutat. Res.*, 31, 347, 1975.
4. Benedict, W. F., Jones, P. A., Laug, W. E., Igel, H. J., and Freeman, A. E., *Nature (London)*, 256, 322, 1975.
5. Benedict, W. F., Rucker, N., Faust, J., and Kouri, R. E., *Cancer Res.*, 35, 857, 1975.
6. Benedict, W. F., *Nature (London)*, 260, 368, 1976.
7. Benedict, W. F., Banerjee, A., Gardner, A., and Jones, P. A., *Cancer Res.*, 37, 2202, 1977.
8. Bertram, J. S. and Heidelberger, C., *Cancer Res.*, 34, 526, 1974.
9. Bertram, J. S., *Cancer Res.*, 37, 514, 1977.
10. Berwald, Y. and Sachs, L., *Nature (London)*, 200, 1182, 1963.
11. Bouck, N. and DiMayorca, G., *Nature (London)*, 264, 722, 1976.
12. Casto, B. C., *Cancer Res.*, 33, 402, 1973.
13. Casto, B. C. and DiPaolo, J. A., *Prog. Med. Virol.*, 16, 1, 1973.
14. Casto, B. C., Pieczynski, W. J., Janosko, N., and DiPaolo, J. A., *Chem. Biol. Interact.* 13, 105, 1976.
15. Casto, B. C., Janosko, N., and DiPaolo, J. A., *Cancer Res.*, 37, 3508, 1977.
16. DiMayorca, G., Greenblatt, M., Trauthen, T., Soller, A., and Giordano, R., *Proc. Natl. Acad. Sci. U. S. A.*, 70, 46, 1973.
17. DiPaolo, J. A., Donovan, P. J., and Nelson, R. L., *J. Natl. Cancer Inst.*, 42, 867, 1969.
18. DiPaolo, J. A., Nelson, R. L., and Donovan, P. J., *Science*, 165, 917, 1969.
19. DiPaolo, J. A., Nelson, R. L., and Donovan, P. J. *Nature (London)*, 235, 278, 1972.
20. DiPaolo, J. A., Takano, K., and Popescu, N. C., *Cancer Res.*, 32, 2686, 1972.
21. DiPaolo, J. A., Nelson, R. L., Donovan, P. J., and Evans, C. H., *Arch. Pathol.*, 95, 380, 1973.

22. DiPaolo, J. A., in *Chemical Carcinogenesis, Part B,* Ts'o, P. O. P. and DiPaolo, J. A., Eds., Marcel Dekker, New York, 1974, 443.
23. DiPaolo, J. A. and Casto, B. C., in *Screening Tests for Chemical Carcinogenesis,* Montesano, R., Bartsch, H., and Tomatis, L., Eds., Publ. No. 12, International Agency for Research on Cancer, Lyon, France, 1976, 415.
24. Doll, R., *Nature (London),* 265, 589, 1977.
25. Dunkel, V. C., in *Screening Tests in Chemical Carcinogenesis,* Montesano, R., Bartsch, H., and Tomatis, L., Eds., Publ. No. 12, International Agency for Research on Cancer, Lyon, France, 1976.
26. Farber, E., *Cancer Res.,* 33, 2537, 1973.
27. Freedman, V. H. and Shin, S., *J. Natl. Cancer Inst.,* 58, 1873, 1977.
28. Freeman, A. E., Weisburger, E. K., Weisburger, J. H., Wolford, R. G., Maryok, J. M., and Heubner, R. J., *J. Natl. Cancer Inst.,* 51, 799, 1973.
29. Freeman, A. E., Igel, H. J., and Price, P. J., *In Vitro,* 11, 107, 1975.
30. Gordon, R. J., Bryan, R. J., Rhim, J. S., Demoise, C., Wotford, R. G., Freeman, A. E., and Heubner, R. J., *Int. J. Cancer,* 12, 223, 1973.
31. Grover, P. L., Sims, P., Huberman, E., Marquardt, H., Kuroki, T., and Heidelberger, C., *Proc. Natl. Acad. Sci. U. S. A.,* 68, 1098, 1971.
32. Haber, D. A., Fox, D. A., Dynan, W. S., and Thilly, W. G., *Cancer Res.,* 37, 1644, 1977.
33. Heidelberger, C., *Adv. Cancer Res.,* 18, 317, 1973.
34. Heidelberger, C., *Annu. Rev. Biochem.,* 44, 79, 1975.
35. Huberman, E. and Sachs, L., *Proc. Natl. Acad. Sci. U. S. A.,* 56, 1123, 1966.
36. Huberman, E., Salzberg, S., and Sachs, L., *Proc. Natl. Acad. Sci. U. S. A.,* 59, 77, 1968.
37. Huberman, E., Kuroki, T., Marquardt, H., Selkirk, J. K., Heidelberger, C., Grover, P. L., and Sims, P., *Cancer Res.,* 32, 1391, 1972.
38. Huberman, E., Donovan, P. J., and DiPaolo, J. A., *J. Natl. Cancer Inst.,* 48, 837, 1972.
39. Huberman, E., Mager, R., and Sachs, L., *Nature (London),* 264, 360, 1976.
40. Ishii, Y., Elliot, J. A., Mishra, N. K., and Lieberman, M. W., *Cancer Res.,* 37, 2023, 1977.
41. Jones, P. A., Benedict, W. F., Baker, M. S., Mondal, S., Rapp, U., and Heidelberger, C., *Cancer Res.,* 36, 101, 1976.
42. Jones, P. A., Baker, M. S., Bertram, J. S., and Benedict, W. F., *Cancer Res.,* 37, 2214, 1977.
43. Kakunaga, T., *Int. J. Cancer,* 12, 463, 1973.
44. Kakunaga, T., *Cancer Res.,* 35, 1637, 1975.
45. Katsuda, H., Takaoka, T., and Yamada, T., in *Chemical Carcinogenesis, Part B,* Ts'o, P. O. P., and DiPaolo, J. A., Eds., Marcel Dekker, New York, 1974, 473.
46. Kier, L. D., Yamasaki, E., and Ames, B. N., *Proc. Natl. Acad. Sci. U. S. A.,* 71, 4159, 1974.
47. Kuroki, T. and Sato, H., *J. Natl. Cancer Inst.,* 41, 53, 1968.
48. Lasne, C., Gentil, A., and Chouroulinkov, I., *Br. J. Cancer,* 35, 722, 1977.
49. Ledinko, N. and Evans, M., *Cancer Res.,* 33, 2936, 1973.
50. Mager, R., Huberman, E., Yang, S. K., Gelboin, H. V., and Sachs, L., *Int. J. Cancer,* 19, 814, 1977.
51. Marquardt, H., in *Screening Tests for Chemical Carcinogens,* Montesano, R., Bartsch, H., and Tomatis, L., Eds., Publ. No. 12, International Agency for Research on Cancer, Lyon, France, 1976, 389.
52. Marquardt, H., Grover, P. L., and Sims, P., *Cancer Res.,* 36, 2059, 1976.
53. Marquardt, H., Philips, F. S., and Sternberg, S. S., *Cancer Res.,* 36, 2065, 1976.
54. Marquardt, H. and Baker, S., *Cancer Lett.* 3, 31, 1977.
55. Mishra, N. K. and DiMayorca, G., *Biochim. Biophys. Acta,* 355, 205, 1974.
56. Mondal, S., Brankow, D. W., and Heidelberger, C., *Cancer Res.,* 36, 2254, 1976.
57. Mondal, S. and Heidelberger, C., *Nature (London),* 260, 710, 1976.
58. Nesnow, S. and Heidelberger, C., *Cancer Res.,* 36, 1801, 1976.
59. Pienta, R. J., Poiley, J. A., and Lebherz, W. B., III, *Int. J. Cancer,* 19, 642, 1977.
60. Price, P. J., Suk, W. A., and Freeman, A. E., *Science,* 177, 1003, 1972.
61. Purchase, I. F. H., Longstaff, E., Ashby, J., Styles, J. A., Anderson, D., LeFevre, P. A., and Westwood, F. R., *Nature (London),* 264, 624, 1976.
62. Reznikoff, C. A., Bertram, J. S., Brankow, D. W., and Heidelberger, C., *Cancer Res.,* 33, 3239, 1973.
63. Reznikoff, C. A., Brankow, D. W., and Heidelberger, C., *Cancer Res.,* 33, 3231, 1973.
64. Rhim, J. S. and Huebner, R. J., *Cancer Res.,* 33, 695, 1973.
65. Rhim, J. S., Park, D. K., Weisburger, E. K., and Weisburger, J. H., *J. Natl. Cancer Inst.,* 52, 1167, 1974.
66. Rhim, J. S., Park, D. K., Arnstein, P., Huebner, R. J., Weisburger, E. K., and Nelson-Rees, W. A., *Nature (London),* 256, 751, 1975.
67. Schoental, R., *Nature (London),* 215, 535, 1967.
68. Stoker, M. and MacPherson, I., *Nature (London),* 203, 1355, 1964.
69. Terzaghi, M. and Little, J. B., *Cancer Res.,* 36, 1367, 1976.
70. Ts'o, P. O. P., *J. Toxicol. Environ. Health,* 2, 1305, 1977
71. Waters, R., Mishra, N., Bouck, N., DiMayorca, G., and Regan, J. D., *Proc. Natl. Acad. Sci. U. S. A.,* 74, 238, 1977.

72. **Weinstein, I. B., Wigler, M., and Stadler, U.**, in *Screening Tests in Chemical Carcinogenesis,* Montesano, R., Bartsch, H., and Tomatis, L., Eds., Publ. No. 12, International Agency for Research on Cancer, Lyon, France, 1976, 355.
73. **Weisburger, J. H.**, *J. Occup. Med.*, 18, 245, 1976.
74. **Wynder, E. L. and Gori, G. B.**, *J. Natl. Cancer Inst.*, 58, 825, 1977.
75. **Pienta, R. J. et al.**, personal communication.

GENETIC ASPECTS OF SHORT-TERM TESTING AND AN APPRAISAL OF SOME IN VIVO MAMMALIAN TEST SYSTEMS

E. R. Soares

TABLE OF CONTENTS

I. Introduction 67

II. Genetic Considerations in Short-Term Testing 67

III. In Vivo Mutagenicity Tests 68
    A. Somatic Cell Assays 69
        1. Micronucleus Test 69
        2. Somatic Cell Cytogenetics 69
        3. Sister Chromatid Exchange 70
        4. Recessive Spot Test 71
    B. Germ Cell Assays 72
        1. Dominant Lethal Assay 72
        2. Heritable Translocation Assay 73
        3. Sperm Morphology 73
        4. Germ Cell Specific Locus Test 74

References 75

I. INTRODUCTION

There are two major areas to address in the course of this discussion. The first concerns the importance of genetics and genetic standardization of experiments to short-term in vivo and in vitro testing for mutagens and carcinogens. The second area will be oriented toward a brief description of some of the in vivo mutagenicity tests presently available and their advantages and disadvantages.

II. GENETIC CONSIDERATIONS IN SHORT-TERM TESTING

From a cursory scan of the numerous procedures presently used in testing for mutagens and carcinogens, it is immediately obvious that the majority of these assays use as end points some form of genetic damage. These end points include measurement of: (1) mutations at individual loci (Ames test; mutagenicity in CHO cells), (2) DNA damage and repair, and (3) clastogenic events including deletions and chromatid and chromosome breakage and repair. It is interesting to note that despite this reliance upon genetic damage or damage to the genetic material, very little emphasis is placed upon standardizing protocols with regard to genetic parameters. That is, little effort is made to control variability inherent to the tissues or experimental organisms used in the "routine" testing procedures.

Variations in the response of animals of different species or strains to environmental insults are well documented.[21,28,44] Numerous studies have shown that a large component of these observed differences is directly correlated with the genotype of the animal.[18,22,35,36] The genetic makeup of an animal determines to a certain extent the response of that animal to particular environmental changes. Included in these changes and in the aforementioned studies are the effects of various toxicants, such as chemical carcinogens and mutagens. It follows that inherent genetic differences among test animals could be of considerable significance to the outcome, evaluation, and interpretation of experimental results from short-term tests.

These latter points are true for both in vitro systems that require the use of animal tissues for activation enzymes, as well as tests which employ whole animals as the test subject. For example, in vitro assays often require the use of activating systems derived from the tissues (liver, lung, etc.) of whole animals. Although the cells or bacteria used in the in vitro tests are genetically standardized, the genotype of the animal from which the tissues are taken is seldom considered. Standardization of the genetic makeup of activating systems (i.e., the genotype of the animal) could be extremely important to the production of accurate and experimentally reproducible results. Unless controlled, widely varying results due to inherent genetic variability between microsomal activation systems within an experiment, between experiments, or among laboratories could be the outcome of otherwise well-controlled studies. The identical rationale would apply to studies using whole animals. Unless the variation introduced by differing genetic backgrounds between test animals is controlled, misleading results could be obtained.

III. IN VIVO MUTAGENICITY TESTS

There are presently dozens of whole animal tests in use for mutagenicity. Each of these tests differs from the others, in choice of animal (e.g., *Drosophila* vs. mouse), in end point measured, or in subtle variations in protocol. Furthermore, each in vivo assay has its distinctive advantages over other tests as well as certain disadvantages. In this section, a few such tests have been selected for discussion, but ones which are considered important, either because of their popularity in testing protocols or because they represent promising innovations to in vivo testing programs. Special emphasis has been placed on those protocols that use the mouse as the test organism. This limitation does not present any serious problems with respect to describing mutagenicity testing in vivo in mammalian species, since for the most part test protocols used in the mouse adequately represent those procedures as they apply to other mammalian species. In addition, an appraisal of the utility of *Drosophila* in mutagenicity testing is included.

Drosophila is unique in that it is the most genetically well-characterized and best-understood whole animal (the house mouse, *Mus musculus*, is a distant second). For this reason, *Drosophila* is especially valuable in studying genetic effects of known and potential mutagens. Sobels[43] has noted that the current "availability of specific tester strains, special markers, useful inversions or other rearrangements makes it possible to test in one and the same experiment for the total spectrum of genetic change, ranging from dominant lethals, and chromosome loss to translocations, recessive lethals, deletions, crossing-over, non-disjunction, and complete and mosaic visible mutations in a whole array of germ cell stages." In addition, Casida[10] has shown that *Drosophila* has the potential to metabolically activate many promutagens. This all-around versatility of *Drosophila* in providing an array of genetic end points and now activating systems, coupled with their short generation time and ease and economy of maintenance, are strong points in favor of using fruit flies for mutagenicity testing. Perhaps the major disadvantage to the exclusive use of *Drosophila* in whole-animal mutagen testing are the problems of extrapolating from data obtained with flies to mammals including humans. Similar problems exist in going from mouse to man. However, the similarities in physiology, reproduction, and gametogenesis (especially oogenesis) make the mouse more attractive for extrapolation. Nevertheless, *Drosophila* remains an extremely useful test organism for early screening of potential mutagens, as well as a tool for studying the genetic effects and mechanism of induction of mutation.

The house mouse ranks second to *Drosophila* with respect to its genetic characterization and standardization. Our knowledge of the genetics of mouse, the availability of genetically standardized stocks and strains, and the economics of maintaining mice, as opposed to rats or rabbits, make the mouse generally the animal of choice in short- and long-term mammalian mutagenicity studies.

In vivo tests of mutagenicity in the mouse can be divided into two categories: (1) those which measure somatic damage and (2) those which measure germ cell damage. It should be noted that of the two, geneticists are primarily concerned with the latter. Mutations occurring in the germ cells (sperm or ova) have the potential of being passed on to and expressed in a subsequent generation. Somatic cell damage, on the other hand, will not be passed on to a future generation, and although it could represent a hazard to the individual bearing said damage, it does not represent a genetic (heritable) hazard to the progeny of that individual. In addition, somatic cell damage may not necessarily be indicative, either qualitatively or quantitatively, of germ cell damage (e.g., increases in peripheral blood micronuclei may not be indicative of chromosomal aberrations in germ cells). Similarly, although somatic cell assays for mutagenic damage can tell us something about the genetic hazard of a compound, they do not provide an insight into the potential heritability of the damage and hence do not allow for estimates of the potential genetic risk of the observed damage.

A. Somatic Cell Assays

1. Micronucleus Test

The micronucleus test was proposed by Heddle[24] as a potentially useful assay for in vivo chromosomal damage. The test involves the microscopic examination of cytological preparations of reticulocytes or polychromatic erythrocytes[49] obtained from the bone marrow, usually the femur, of treated and control animals. The slides are normally prepared between 12 hr and 4 days following treatment and subsequently scored for the presence of micronuclei (acentromeric chromosome fragments). Increases in the frequency of micronuclei among treated animals over control animal values is taken as an indication of treatment-induced genetic damage (i.e., chromosomal aberrations). The techniques of identifying and scoring micronuclei are relatively easy to master and the technique does not require metaphase analysis of chromosomal anomalies. In addition, the test itself is rapid and economical. There are some problems of variability in frequencies of micronuclei, but these can be dealt with by using appropriate sample sizes (numbers of animals as well as cells counted) and through application of valid statistical procedures that take into account variation about the mean.

A serious problem with the micronucleus test concerns the interpretation of the data. As noted earlier, the assay scores for chromosome fragments in the cytoplasm of interphase cells. The problem in interpretation stems from the fact that although the end point, micronuclei, is representative of genetic damage, the actual cause of formation of micronuclei may not be a direct result of treatment-induced genetic damage. The appearance of micronuclei provides only indirect evidence of genetic damage. The possible cause of micronuclei could be the result of a physiological response of a cell to a toxicant and not the result of a direct interaction of the mutagen with the genetic material. Consequently, and as Schmid[40] points out, unexpected test results produced by the micronucleus test should be verified by other in vivo or in vitro studies.

2. Somatic Cell Cytogenetics

As is indicated in the name, this assay involves cytogenetic analysis of chromosome and chromatid aberrations in preparations of metaphase chromosomes of somatic cells (usually bone marrow or peripheral leukocytes). This assay differs from the micronucleus test in that it entails direct observation and categorization, as to type, of the specific aberrations.

In general, the experimental design used in this test follows that used in the micronucleus test. Animals, males or females, are treated with either the test compound or a control compound. At varying times following injection, tissues, either blood or bone marrow, are removed from the animal. In the case of bone marrow, the tissue may be incubated for 30 min in a buffer solution, with colcemid added, before slides are made. In the case of peripheral blood, a more lengthy culture period of 48 to 72 hr[46] is necessary before slides can be made. Once prepared, slides are screened for cells in metaphase. Individual metaphases are examined for the presence of aberrations including breaks, gaps, deletions, translocations, and ploidy changes. Increases in aberrations in one or more of these categories are considered indicative of mutagenic activity.

Perhaps the most important advantage of this test is that it provides for direct observation of chromosomal or chromatid aberrations and, unlike the micronucleus test, allows for characterization of the chromosomal damage. A second major advantage to this assay stems from the fact that peripheral blood may be used as a source of leukocytes. Since one is not limited to bone marrow and since blood can be sampled relatively easily, this assay is applicable to the observation of human chromosomes.[4,34] Also, this test can be accomplished fairly rapidly and requires no special breeding protocols or elaborate experimental designs.

Considerable caution should be exercised in interpreting data from this assay. It is important to note that this test is primarily designed to detect those chemicals that cause chromosome breakage and is not an efficient methodology for screening single gene mutations. Also, as with the micronucleus test, it is possible that the observed aberrations are not the result of the direct action of the compound on the chromosomes. This is especially important to human studies where less experimental control is possible. It is well known that such biological factors as viral infections can be responsible for considerable increases in chromosomal aberrations.[13] If aberrations are detected using this assay, and if they are the result of the action of a mutagen, it must be remembered that these lesions have occurred in somatic cells and may not be at all representative of what has occurred in the germ cells.[6] Consequently, any estimates of genetic risk to humans based upon data from this test should be made with the above restrictions and problems in mind.

One further problem with the somatic cell cytogenetic assay is that of mastering the technique of reading and interpreting the metaphase chromosomes. This is a task that requires cytogenetic training and is considerably more laborious than simply scoring acentromeric fragments. This is especially true of mouse chromosomes where all 20 chromosomes are acrocentric or telocentric with only relatively minor differences in size. The same problem of technique is evident with human chromosomes, but is somewhat alleviated by a more easily categorized karyotype.

3. Sister Chromatid Exchange

Identification of sister chromatid exchanges (SCE) were originally described in a paper by Taylor et al.,[45] in which autoradiographic techniques were employed to identify reciprocal exchanges between chromatids. Recently, Latt[30-32] has utilized BrdU incorporation into replicating chromosomes and special staining techniques to identify SCEs. Regardless of methodology, the technique is designed to afford the identification of exchanges (apparently reciprocal) between homologous segments of sister chromatids. In theory, these exchanges result from chromatid (DNA) breakage and subsequent reunion.[29]

In general, the methodology requires the exposure of the cells of interest to BrdU during two consecutive replications. During replication, BrdU is differentially incorporated into both strands of one chromatid and just one strand of the homologous or sister chromatid. Subsequent staining of chromosomes from these cells using the stain 33258 Hoechst allows identification of sister chromatids through their differential fluorescence. The doubly substituted chromatid (DNA) appears as a dully fluorescence chromatid, while the singly substituted DNA fluoresces brightly. This differential fluorescence between sister chromatids allows for the identification of SCEs.

Until just recently, this technique was used exclusively for the identification of SCEs in in vitro preparations. Bloom and Hsu[5] using the chick embryo and Allen and Latt[1] using the mouse have demonstrated that the technique is applicable to in vivo studies of SCEs. Allen's work deals with the identification of SCEs in the germ cells of mice and opens the door to a possible new test system for mutagens. Such an assay would make feasible direct comparisons, using exactly the same end points, of the effects of a mutagen in in vitro cell systems and in vivo. Allen and Latt[1] have extended this possibility to in vivo studies of SCE formation in somatic and meiotic cells of the mouse.

One problem of in vivo SCE analysis involves the administration of BrdU to the animals. Allen et al.[3] have suggested the use of subcutaneous implantation of BrdU in tablet form as an alternative to the laborious procedures involving 14 hourly injections. A second area of concern is with the interpretation of the significance of SCE formation. It is clear from the work cited above that treatment with clastogens causes an increase in SCEs and further that these increases are probably the result of DNA breakage. However, because BrdU is itself mutagenic, it is not possible to determine the spontaneous rate of SCE formation or determine whether active compounds are just enhancing the effects of BrdU. Additional work in evaluating the genetic significance and consequences of SCE formation is necessary.

The theoretical origin of SCEs is through DNA breakage. Consequently, a compound that produces a genetic alteration by another mechanism might not induce SCEs. This is a point that should be considered especially in reviewing data produced in testing suspect mutagens whose mechanism of action is unknown.

4. Recessive Spot Test

Most in vivo, somatic cell mutagenicity test systems presently available are designed to detect mutagenic events which are characterized by either chromatid or chromosomal aberrations (micronucleus, somatic cell cytogenetics, and sister chromatid exchange). A possible alternative to these assays is the somatic cell recessive spot test.[37] This assay was originally proposed as a means of evaluating the mutagenic effects of radiation. More recently, the utility of this test as a screen for chemical mutagens has been recognized and some effort has been placed upon the validation of this system with known mutagens or carcinogens.[17,38]

This assay is designed to detect induced mutations at several coat color loci in selected strains of mice. The spot test is unique to other short-term tests in that it is specifically designed to detect single gene lesions as opposed to chromosomal aberrations. The target cells in this assay are somatic, but more specifically are the embryonic, pre- or early migratory neural crest cells. These cells are responsible for the production of melanocytes that give the characteristic coat color as it is determined by the genotype of the animal. The coat color genes used as markers in this assay are well defined, and mutations of these genes are known to cause reasonably predictable alterations in color. Therefore, mutations at these loci in the neural crest cells result in alterations of coat color which appear as spots at the final sight of migration of the mutant cell(s).

In running this assay, crosses are made between two genetically standardized stocks:

$$\text{C57BL/6 } \frac{a}{a}; \frac{+^b}{+^b}; \frac{+c^{ch}, +^p}{+c^{ch}, +^p}; \frac{+^d, +^{se}}{+^d, +^{se}}; \frac{+^s}{+^s} \text{ Black} \quad (1)$$

$$\text{T-Stock } \frac{a}{a}; \frac{b}{b}; \frac{c^{ch} \; p}{c^{ch} \; p}; \frac{d, se}{d, se}; \frac{s}{s} \text{ White} \quad (2)$$

Females are then treated with the test compound generally at 10.25 days following conception and are allowed to bring their litters to parturition and weaning. At 3 to 5 weeks of age, all offspring are scored for coat color spots that are considered indicative of induced mutation at one or more of the marker loci.

The recessive spot test has a number of major advantages. It is a relatively rapid assay and is economically reasonable both from the point of view of cost and sample size. Since the target cells in the spot test are embryonic neural crest cells which number approximately 150 to 200 at the time of treatment, relatively few pregnant females are required to provide an experimentally as well as a statistically adequate sample. Since the sample size is reasonably small, the cost is not overly restrictive. The technique is designed to detect gene mutations and consequently should be able to identify nonclastogenic mutagens as well as clastogens which produce minor aberrations (deletions) of the marker loci. In addition, the test measures mutation in the fetus, which provides information as to whether a compound can cross the placental blood barrier.

As has been noted by Russell,[37] there are some problems in identifying true recessive spots, as opposed to spots that are not the result of mutation. Misclassification of this kind could lead to spurious results. However, mastery of the technique of distinguishing mutational lesions from "artifacts" requires only practice and some basic knowledge of the biology and genetics of the mouse. A second problem that may or may not be serious, depending upon point of view and intentions, stems from the fact that the test deals with somatic as opposed to germinal cells. Consequently, estimates of mutations rates and/or genetic risk in this test may not be representative of germ cell mutagenesis. This difficulty has been considered by Russell[37] and recent work indicates that the somatic cell specific locus mutation rate may not be terribly different from the germ cell rate. Additional work in this respect needs to be done and in fact is underway.

One last point regarding time of treatment should be noted. Any variation in time of exposure will result in changes in the number of neural crest cells (i.e., sample size) being treated. Consequently, treatment timing must be controlled precisely so as to avoid treating too few cells (early) or too many cells (late). The former would result in a lower probability of detecting mutations because of decreased sample size. The latter would result in a larger sample size but would give mutant spots too small to detect.

B. Germ Cell Assays

1. Dominant Lethal Assay

The dominant lethal assay (DLA) is one of the most widely used methodologies for determining the mutagenicity of a compound or radiation in vivo in mice. Although minor variations are not unusual, the general design for the dominant lethal test has been fairly well standardized. Male mice are treated with the suspect mutagen, while a second group of males is sham treated and used as a control. Subsequent to treatment (time may vary, but usually 24 to 48 hr after treatment) each male is paired with one or more female mice. These matings are maintained until vaginal plugs are found in the females or for about 4 to 7 days, at which time the females are replaced with additional virgin females. Mated females are caged separately. This scheme is repeated over a 6- 8-week period in order that all stages of the spermatogenic cycle are tested for induced mutagenic damage. Females are killed 17 days after initial exposure to a male and their uteri are examined for the presence of living and dead embryos (implantations). In evaluating their data, investigators generally make comparisons between groups with respect to total living and dead implantations. In general, an increase in dead implantations in the treated group, which may or may not be accompanied by a decrease in total implantations, is considered indicative of induced dominant lethals.

The DLA has several important advantages. It is a reasonably rapid assay (10 weeks) and is not economically prohibitive. It measures the effects of mutagens on germ cells and measures varying effects of mutagens on different germ cell stages. It can be used in assaying the effects of compounds on females, but this presents some interpretative problems with respect to biological and physiological differences between males and females.[20] Although there has been some question in the past as to the genetic basis for dominant lethals, recent data of Burki and Sheridan[8] indicate dominant lethals are closely tied to clastogenic events. Although the dominant lethal events could be single gene mutations, it appears that they are more likely the result of chromosomal aberrations.

In addition to its obvious advantages, the DLA has several disadvantages. Included among the disadvantages of the DLA are

1. From the data of Burki it appears that this assay measures only clastogenic events and would not detect nonclastogenic agents.
2. It is a relatively (as compared to the translocation and the germ cell specific locus assays) insensitive test. This is a result of the high and variable frequencies of spontaneous lethals which, in turn, result in problems of experimental design (sample size) and statistical analysis. With respect to sample size, it is virtually impossible to recommend a "good" sample size because in general, the weaker the mutagen (i.e., the smaller the effect that is to be detected), the larger the sample size required. Statistical problems, on the other hand, have been fairly well worked out[23] and adequate yet simple procedures are available.
3. The DLA, by definition, provides no means for further genetic analyses. However, if the data of Burki are accurate, the DLA could by association provide an estimate of the clastogenic effects of a compound.
4. The DLA has been referred to as an irrelevant test, since by definition dominant lethals pose no genetic hazard to future generations. It should be noted that fetal death or loss by miscarriage is not irrelevant to humans and must be considered in evaluating risks to humans.

2. Heritable Translocation Assay

As is indicated by the name, the heritable translocation assay (HTA) is designed to detect chromosomal aberrations in the form of translocations which have occurred in the germ cells of test individuals and which are subsequently passed on to the progeny of those individuals.

Like most other in vivo assays, there are several variations in designs that are used to screen for induced translocations.[21,33] In general, male mice are treated with the suspect agent and subsequently paired with females. An equal number of sham-treated controls may not be necessary, since spontaneous translocations are extremely rare. Historical data from other experiments in a laboratory and/or with the same strain of mice, along with a small number of controls, would probably be a valid alternative (but this method is not problem-free). Matings of treated males may be continued through the spermatogenic cycle or may be terminated at any point during the cycle. Mated females are allowed to go to parturition and wean their pups. Male offspring (females are usually discarded) are raised to sexual maturity and tested for partial sterility or sterility (both phenomena are possible consequences of heterozygosity for a reciprocal translocation). Here, each male is mated to three virgin females for seven nights at which time the females are replaced and a second mating period is established. All the females are killed 17 days following initial exposure to a male and are examined for the presence of dead and living embryos. A significant decrease in live embryos (theoretically 50%) or the absence of any embryos is considered indicative of semisterility or sterility, respectively, that is of a translocation. Cytogenetic analysis of meiotic cells from suspect translocation bearers would then be done to confirm the presence of a translocation.[16]

Perhaps the major advantage of the HTA is that it is capable of detecting *true genetic* lesions of possible significance to subsequent generations. In addition, the design used in the HTA allows for the maintenance and additional study of translocation-bearing animals and consequently the genetic effects of the induced aberrations. The HTA appears to be considerably more sensitive[19] than the DLA, without being overly expensive.

The heritable translocation test does suffer from some important disadvantages. It is designed to detect clastogenic events and of these only those that result in translocations which in turn cause aberrant fertility on the part of the "mutant" bearers. Inversions, small deletions, Robertsonian translocations, or perhaps small terminal translocations could go undetected. Experimental sample size could also be a problem. In the absence of a large historical control, reliable statistical evaluation of the effects of a weakly mutagenic compound would necessitate a prohibitively large test sample. With an historical control, calculations of confidence intervals could be made which might allow for the use of a smaller sample. One final drawback to this assay involves the cytogenetic confirmation of the translocation. Such a cytogenetic verification can be a technically rigorous procedure and one that requires, with few exceptions, specialized training in the area of cytogenetics.

3. Sperm Morphology

The sperm morphology assay (SMA) is a relatively new in vivo assay for the effects of chemicals on the germ cells of male mice.[7] The endpoint in the SMA is the percent increase of morphologically aberrant sperm in the treated series as compared to the control. Sperm head shape and tail attachment are the two parameters that are considered. Tailless sperm are excluded from the counts. The experimental design for the SMA is straightforward, requires no long-term breeding protocols, and remains essentially unchanged regardless of the type of compound used or the route of administration. Individual male mice are treated with the test compound. At varying times following treatment (2, 4, 6, and 8 weeks), males are sacrificed and sperm samples are collected from the vas deferens and/or the epididymis of each male. Samples are placed in saline and stained for approximately 30 min with a 1% eosin solution, following which slides are prepared using standard blood smear techniques. Counts of normal and abnormal sperm are made at approximately 800×.

There are a number of advantages to the SMA over and above the fact that it uses direct measurements on germ cells. The technique of distinguishing between normal and abnormal sperm can be mastered with practice and should result in consistent and reproducible data. Recent studies[25,48] have shown the SMA to be rapid, reliable, and extremely sensitive. It is especially important to note that this assay is capable of detecting the effects of mutagens in germ cell stages (spermatocytes, spermatogonia) in which other assays (the dominant lethal, heritable translocation test) have given no or only negligible indications of mutagenic effects at similar doses. The study by Heddle and Bruce[25] indicated that the SMA was far superior to the micronucleus test and compared well with the Ames test in correlating with known mutagens/carcinogens and known nonmutagens/noncarcinogens.

One major concern with the SMA is the seemingly subjective categorization of sperm as either normal or abnormal. Such subjectivity does exist, but can be controlled within a laboratory by simply limiting slide reading to certain individuals and by standardizing sperm types within particular categories. The problem of subjectivity between laboratories is less amenable to control and would probably be best alleviated through statistical considerations. Another major problem is the interpretation of the data. Although the cells being analyzed are germ cells (sperm), mutagen-induced increases in abnormalities do not necessarily mean that the damage is genetic in origin. Further, it is not possible to isolate individual aberrant sperm for additional *genetic* testing. However, in light of the data cited above, regarding germ cell stages, one would expect some proportion of the induced damage to be genetic, and in fact Hugenholtz and Bruce[26] have shown the induction and transmission to the offspring of elevated levels of abnormal sperm. On the other hand, Wyrobek et al.[47] have shown that increases in sperm abnormalities are not associated with chromosome abnormalities. Although it has been reported that mutagens increase the level of sperm abnormalities, the actual basis for this increase is unclear and additional work is necessary before the genetic component of induced sperm abnormalities can be distinguished from other causes.

As with many other in vivo tests, sample size and proper statistical treatment of the data are extremely important to the SMA. Soares et al.[42] have addressed these concerns and note that neither are prohibitively burdensome. In addition, they have focused attention on the problem of "outliers" and have noted that it is reasonable to expect individual animals to show high levels of abnormalities that are apparently unrelated to treatment. However, such outliers are easily recognized and can be eliminated as a source of error by applying relatively simple statistical procedures.

4. Germ Cell Specific Locus Test

The specific locus test (SLT) was originally developed by Russell[39] in experiments involving the effects of radiation. A second SLT using the mouse was developed by Carter et al.[9] Unlike all of the above tests, the SLTs are designed to detect single gene mutations arising in the germ cells of mice. The SLT has the capability of detecting recessive mutations at specific loci, and dominant visibles, as well as certain deletions and translocations.[41]

Experimentally, the test[39] requires two strains of mice one of which is a multiple recessive stock for the seven specific loci (T-stock) and one of which is a homozygous wild type for the seven specific test loci. The wild-type animals (usually males) are exposed to the test substance and subsequently mated to the T-stock animals. Matings can be carried out at various times after exposure to allow for testing of the different stages of spermatogenesis. F_1 offspring from these matings will exhibit the wild-type phenotype unless a mutation has occurred at one of the loci, in which case an altered phenotype will be apparent. Any such divergence from the wild type is considered indicative of a putative mutation. Following initial identification of the "mutant" animal, additional tests (progeny testing, allelism tests) are conducted to verify the heritability of the variant and to determine more exactly the nature of the mutation.

This test or variations of the original SLT has been used to study the effects of irradiation for some 25 years and recently[11,12,14,15] to study the mutagenic effects of chemicals. Currently, it is the only available methodology for identifying specific locus mutations in the mouse. Included among its advantages are the following:

1. It is a genetic test in the true sense, since it deals with transmissable mutations.

2. By simply observing F_1 offspring, one can detect the occurrence of both recessive and dominant visible mutations.
3. Because F_1 progeny, including mutant bearers, are maintained, the SLT allows for additional genetic testing of putative mutations.
4. Because of the nature of the design used in the SLT, this test can be combined with other tests such as the DLA or translocation test.

Among its disadvantages are:

1. It requires almost prohibitively large sample sizes. To this point, it is important to note that a larger historical control is available and could be used in the manner described earlier under translocations, as a means of reducing sample size.
2. The SLT has the capability of detecting only those recessive mutations that occur at the specific loci and which result in a phenotypic change.
3. Although a phenotypic change and subsequent genetic testing give clear evidence of a genetic effect, the biochemical or molecular genetic cause of said effect is difficult to elucidate absolutely.
4. Lastly, mutations at the specific loci which are deleterious in the heterozygous state may be eliminated prior to observation.[41]

It is not the purpose of this paper to cover in detail all the available in vivo mammalian assays for mutagenicity, but only to consider those tests which are widely used or which, in the opinion of the author, represent practical innovations to the field of mutagenicity testing. Consequently, protocols that would test for the induction of, for example, recessive lethals have been excluded for the present time. Nevertheless, it is important to realize that considerable work is being done in this area, and with the anticipated progress in this field, additional assays may become quite practical as part of a testing battery.

REFERENCES

1. **Allen, J. W. and Latt, S. A.**, Analysis of sister chromatid exchange formation *in vivo* in mouse spermatogonia as a new test for environmental mutagens, *Nature (London)* 260, 449, 1976.
2. **Allen, J. W. and Latt, S. A.**, *In vivo* BrdU-33258 hoechst analysis of DNA replication kinetics and sister chromatid exchange formation in mouse somatic and meiotic cells, *Chromosoma*, 58, 325, 1976.
3. **Allen, J. W., Shuler, C. F., Mendos, R. W., and Latt, S. A.**, A simplified technique for *in vivo* analysis of sister-chromatid exchanges using 5-bromodeoxyuridine tablets, *Cytogenet. Cell Genet.*, 18, 231, 1977.
4. **Arakaki, D. T.**, Clinical cytogenetics in Hawaii — an introduction to chromosome methodology, *Cytogenetics*, 34, 241, 1975.
5. **Bloom, S. E. and Hsu, T. C.**, Differential fluorescence of sister chromatids in chicken embryos exposed to 5-bromodeoxyuridine, *Chromosoma*, 51, 261, 1975.
6. **Brewen, J. G. and Preston, R. J.**, Chromosome aberrations as a measure of mutagenesis *in vivo* and *in vitro* and in somatic and germ cells, *Environmental Health Perspect.*, 6, 157, 1973.
7. **Bruce, W. R., Furrer, R., and Wyrobek, A. J.**, Abnormalities in the shape of murine sperm after acute testicular X-irradiation, *Muta. Res.*, 23, 381, 1974.
8. **Burki, K. and Sheridan, W.**, Expression of TEM-induced damage to postmeiotic stages of spermatogenesis of the mouse during early embryogenesis. I. Investigations with *in vitro* embryo culture, *Mutat. Res.*, 49, 259, 1978.
9. **Carter, T. C., Lyon, M. F., and Phillips, R. J. S.**, Induction of mutations in mice by chronic gamma irradiation, *Br. J. Radiol.*, 29, 106, 1956.
10. **Casida, J. E.**, Insect microsomes and insecticide chemical oxidations, in *Microsome and Drug Oxidation*, Gillette, J. R., et al., Eds., Academic Press, New York, 1969, 517.
11. **Cattanach, B. M.**, Specific locus mutations in mice, *Chemical Mutagens: Principles and Methods for Their Detection*, Vol. 2, Hollander, A., Ed., Plenum Press, New York, 1971, 535.
12. **Cattanach, B. M.**, Chemically induced mutations in mice, *Mutat. Res.*, 3, 346, 1966.
13. **Cooper, J. E. K., Yohn, D. S., and Stitch, H. F.**, Viruses and mammalian chromosomes. X. Comparative studies of the chromosome damage induced by human and simian adnoviruses, *Exp. Cell Res.*, 53, 225, 1968.

14. Ehling, U. H., The multiple loci method, in *Chemical Mutagenesis in Mammals and Man*, Vogel, F. and Rohrborn, G., Eds., Springer-Verlag, Berlin, 1970, 156.
15. Ehling, U. H., Differential spermatogenic response of mice to the induction of mutations by antineoplastic drug, *Mutat. Res.*, 26, 285, 1974.
16. Evans, E. P., Brechken, G., and Ford, C. E., An air-drying method for meiotic preparations from mammalian testes, *Cytogenetics*, 3, 289, 1964.
17. Fahrig, R., A mammalian spot test. Induction of genetic alterations in pigment cells of mouse embryos with X-rays and chemical mutagens, *Mol. Gen. Genet.*, 138, 309, 1975.
18. Festing, M. F. W., A case for using inbred strains of laboratory animals in evaluating the safety of drugs, *Food Cosmet. Toxicol.*, 13, 369, 1975.
19. Generoso, W. M., Russell, W. L., Huff, W. W., and Stout, S. K., Effect of dose on the induction of dominant lethal mutations and heritable translocations with ethylmethanesulfonate in male mice, *Genetics*, 77, 741, 1974.
20. Generoso, W. M., Chemical induction of dominant lethals in female mice, *Genetics*, 61, 461, 1969.
21. Generoso, W. M., Evaluation of chromosome aberration effects of chemicals on mouse germ cells, *Environ. Health Perspect.*, 6, 13, 1973.
22. Green, E. L. and Meier, H., Use of laboratory animals for the analysis of genetic influences upon drug toxicity, *Ann. N.Y. Acad. Sci.*, 123, 295, 1965.
23. Haseman, J. K. and Soares, E. R., The distribution of fetal death in control mice and its implications on statistical tests for dominant lethal effects, *Mutat. Res.*, 41, 277, 1976.
24. Heddle, J. A., A rapid *in vivo* test for chromosomal damage, *Mutat. Res.*, 18, 187, 1973.
25. Heddle, J. A. and Bruce, W. R., Comparisons of tests for mutagenicity or carcinogenicity using assays of sperm abnormalities, formation of micronuclei, and mutations in *Salmonella*, *Cold Spring Harbor Symp. Origins Hum. Cancer*, 4, 1549, 1977.
26. Hugenholtz, A. P. and Bruce, W. R., Induction and transmission of elevated levels of abnormally shaped murine sperm, *Can. J. Genet. Cytol.*, 18, 564, 1976.
27. Kanady, N. J. and Smith, W. R., Variations in the interferogenic responses of two strains of mice, *Proc. Soc. Exp. Biol. Med.*, 141, 794, 1972.
28. Kouri, R. E., Salerno, R. A., and Whitmire, C. E., Relationships between aryl hydrocarbon hydroxylase inducibility and sensitivity to chemically induced sarcomas in various strains of mice, *J. Natl. Cancer Inst.*, 50, 363, 1973.
29. Latt, S. A., Sister chromatid exchange, *Genetics,* Mammalian Mutagenicity Workshop, Sheridan, W. and Roderick, T., Eds., Jackson Laboratory, Bar Harbor, Maine, 1976.
30. Latt, S. A., Microfluorometric detection of DNA replication in human metaphase chromosomes, *Proc. Natl. Acad. Sci. U.S.A.*, 70, 3395, 1973.
31. Latt, S. A., Localization of sister chromatid exchanges in human chromosome, *Science*, 185, 74, 1974.
32. Latt, S. A., Microfluorometric analysis of DNA replication kinetics and sister chromatid exchanges in human chromosomes, *J. Histochem. Cytochem.*, 22, 478, 1974b.
33. Leonard, A., Tests for heritable translocations in male mammals, *Mutat. Res.*, 31, 291, 1975.
34. Moorehead, P. S., Nowell, P. C., and Mellman, W. J., Chromosome preparations of leukocytes cultured from human peripheral blood, *Exp. Cell Res.*, 20, 613, 1960.
35. Niebert, D. W. and Felton, J. S., Importance of genetic factors influencing the metabolism of foreign compounds, *Fed. Proc. Fed. Am. Soc. Exp. Biol.*, 35, 1133, 1976.
36. Oesch, F., Norris, N., Daly, J. W., Gielen, J. E., and Niebert, D. W., Genetic expression of the induction of epoxide hydrase and aryl hydrocarbon hydroxylase activities in the mouse by phenobarbital or 3-methylcholanthrene, *Mol. Pharmacol.*, 9, 692, 1973.
37. Russell, L. B., Radiation induced presumed somatic mutations in the house mouse, *Genetics*, 42(2), 161, 1957.
38. Russell, L. B., *Invivo* somatic mutation systems in the mouse, *Genetics*, Mammalian Mutagenicity Workshop, Sheridan, W. and Roderick, T., Eds., The Jackson Laboratory, Bar Harbor, Maine, 1976.
39. Russell, W. L., X-ray induced mutations in mice, *Cold Spring Harbor Symp. Quant. Biol.*, 16, 327, 1951.
40. Schmid, W., The micronucleus test, *Mutat. Res.*, 31, 9, 1975.
41. Searle, A. G., The specific locus test in the mouse, *Mutat. Res.*, 31, 277, 1975.
42. Soares, E. R., Haseman, J. K., and Segall, M., Triethylenemelamine induced effects on sperm morphology in mice — some statistical considerations, *Mutat. Res.*, in press.
43. Sobels, F. H., The advantages of drosophila for mutation studies, *Mutat. Res.*, 26, 277, 1974.
44. Storer, J. B., On the relationship between genetic and somatic sensitivity to radiation damage in inbred mouse strains, *Radiat. Res.*, 31, 699, 1967.
45. Taylor, J. H., Woods, P. S., and Hughes, W. L., The organization and duplication of chromosomes as revealed by autoradiographic studies using tritium-labelled thymidine, *Proc. Natl. Acad. Sci. U.S.A.*, 43, 122, 1957.
46. Triman, K., Davisson, M., and Roderick, T., A method for preparing chromosomes from peripheral blood in the mouse, *Cytogenet. Cell Genet.*, 15, 166, 1975.
47. Wyrobek, A. J., Heddle, J. A., and Bruce, W. R., Chromosomal abnormalities and the morphology of mouse sperm heads, *Can. J. Genet. Cytol.*, 17, 675, 1975.
48. Wyrobek, A. J. and Bruce, W. R., Chemical induction of sperm abnormalities in mice, *Proc. Natl. Acad. Sci. U.S.A.*, 72, 4425, 1975.
49. Von Ledebur, M. and Schmid, W., The micronucleus test, methodological aspects, *Mutat. Res.*, 19, 109, 1973.

THE USEFULNESS OF DNA DAMAGE AND REPAIR ASSAYS FOR PREDICTING CARCINOGENIC POTENTIAL OF CHEMICALS

J. A. Swenberg and G. L. Petzold

TABLE OF CONTENTS

I. Introduction 77

II. Status of Validation Studies 79
    A. In Vitro Studies...79
    B. In Vivo Studies 81

References 85

I. INTRODUCTION

Increased public and regulatory concern over the safety of chemicals entering the environment has been manifested by additional requirements for carcinogenicity and chronic toxicity testing. The fact that there are not enough qualified toxicologists and pathologists, research facilities, or financial resources to run such studies on all compounds has led to the rapid proliferation of new "short-term tests". The advent of these assays has brought both hope and confusion to scientists, industry, regulatory officials, and consumers. This review will attempt to critically assess the state of the art of one such group of short-term tests, namely, those tests concerned with DNA damage and repair. More specifically, it will discuss mammalian DNA damage and repair assays which have been evaluated against a broad enough spectrum of chemicals to allow some conclusions to be drawn on their ability to predict carcinogenic or mutagenic potential.

We are greatly indebted to Dr. Ann Mitchell of Stanford Research Institute, Dr. Hans Stich of the University of British Columbia, and Dr. Gary Williams of the Naylor-Dana Institute for Disease Prevention, who provided recent published and unpublished data for this compilation.

DNA damage and repair assays fall into three major categories: in vitro DNA damage, in vitro repair, and in vivo DNA damage. It must be understood that these assays have emerged because they correlate with carcinogenic and mutagenic potential, not because the events they measure necessarily represent *the* critical event in carcinogenesis. So that we all have a basic understanding of how these assays relate to the carcinogenic process, two schematics have been included. The first is a simplified overview of the carcinogenic process (Figure 1). Basically, carcinogenesis has been divided into three major stages: initiation, the latency period, and the appearance of clinical neoplasia. Discussions today will limit themselves to the process of initiation and the subsequent repair (or lack of repair) of DNA damage. As you can readily see, other factors can affect the final outcome, i.e., the appearance of clinical neoplasia. Initiation is usually thought of as an irreversible event that programs a cell to undergo neoplastic transformation at some later date. While this statement is true for those cells that

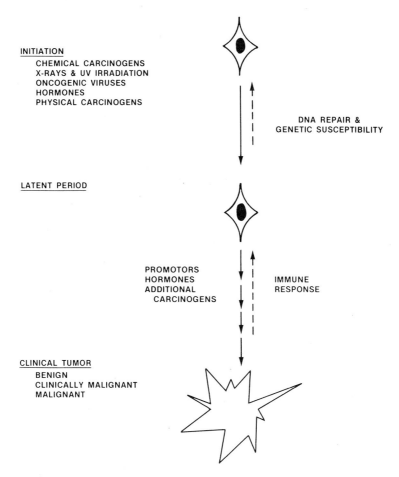

FIGURE 1. The carcinogenic process.

eventually become transformed, it is likely that only a small minority of the cells exposed to carcinogens undergo permanent genetic damage resulting in transformation.[7] Instead, most cells undergo DNA repair or die from cytotoxicity.

Some of the complexities affecting initiation and early events such as DNA repair are illustrated in Figure 2. First, most chemical carcinogens require some degree of metabolic activation to become electrophilic and covalently bind to DNA.[21] During this process, an equally important series of detoxification reactions can take place decreasing the availability of electrophils, thus decreasing the extent of DNA damage. Covalent binding to DNA occurs in many forms. One of the most prominent types of DNA damage is alkylation, leading to the formation of unstable bases such as 7-alkylguanine. These bases can undergo spontaneous hydrolysis, generating apurinic sites that result in template inactivation and subsequent cell death.[33] Apurinic sites can also be generated by the action of N-glycosidase on stable alkylated bases such as O^6-alkylguanine.[13,20] Apurinic and apyrimidinic sites are readily attacked by endonucleases, forming single-strand breaks in the DNA. These can undergo DNA repair that is error free (recovery) or error prone (mutation, cell death, or inconsequential). Stable DNA damage can also persist for long periods of time.[14] Persistent DNA damage can have little effect, cause cytotoxicity, or lead to mutation. Probably the best studied example of this latter event is that of O^6-alkylguanine. This altered base causes mispairing during DNA replication, leading to a permanent and heritable change in the genome of daughter cells.[7] Persistence of O^6-alkylguanine, due to organ-specific repair deficiencies, appears to correlate with the site of tumor formation.[10,15,23]

The ability of chemicals to induce DNA dam-

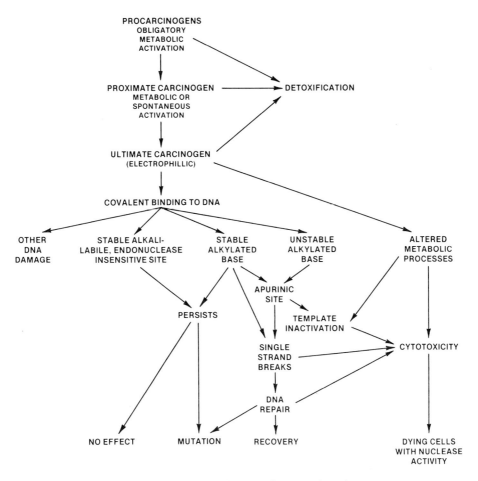

FIGURE 2. Pathways for DNA damage and repair.

age has been studied by three major techniques: alkaline sucrose gradients (ASG),[2-6,12,18,38] alkaline elution (AE),[16,17,24,31] and DNA repair.[1,2,25,28,29,34,35] The first two assays will detect single-strand breaks and alkali-labile lesions, such as apurinic sites. DNA repair assays detect the incorporation of new bases into DNA that is undergoing repair synthesis. In addition to these desirable aspects, ASG and AE techniques will detect single-strand breaks generated by degradative nucleases in dead or dying cells.[37] In vitro systems employing ASG or AE should provide evidence of cell viability at the time of sampling. Such cell viability is not necessarily the same as clonogenic potential, since cells can be damaged to the extent that they are unable to undergo cell division, yet still be alive at the time of assay. Compounds should be evaluated in in vivo ASG or AE assays at non-necrotizing dose levels. To insure this, histological monitoring of tissues should routinely be done. Likewise, DNA repair assays must be controlled to eliminate scheduled DNA synthesis.

II. STATUS OF VALIDATION STUDIES

A. In Vitro Systems

Validation studies using a broad range of chemical carcinogens and noncarcinogens have been run in several in vitro DNA damage and repair assays. Results from these studies are summarized in Table 1. The numbers listed under each investigator represent the number of compounds yielding correct results per total number of compounds tested, assuming that carcinogens cause DNA damage and repair, while noncarcinogens do not. Approximately 75% of the chemicals were carcinogens and 25% noncarcinogens.

Two or more investigators evaluated 52

TABLE 1
DNA Damage and Repair Assays for Predicting Carcinogenic Potential

Chemical class	Total no. chemicals evaluated	DNA repair (UDS)				Alkaline elution, Swenberg[31,32]
		Mitchell[22,28]	Stich[25,29,30]	Williams[34-36]	Ahmed[1]	
Aromatic amines etc.	42	9/12	22/26	9/10	NT[a]	11/12
Alkyl halides etc.	25	3/4	14/15	0/4	10/11	3/6
Polycyclic aromatics	22	5/5	15/16	5/6	NT	6/6
Esters, epoxides, carbamates etc.	23	9/9	5/5	1/1	NT	5/5
Nitroaromatic and heterocycles	45	5/6	30/31	NT	NT	10/12
Miscellaneous aliphatics and aromatics	19	1/1	9/10	2/2	NT	7/9
Nitrosamines etc.	12	5/5	6/6	2/2	NT	7/7
Fungal toxins and antibiotics	12	1/1	6/8	4/4	NT	4/4
Azo Dyes and Miscellaneous	19	10/11	2/3	5/5	NT	6/7
Metallic salts	8	3/5	5/5	NT	NT	NT
Total	262/293 89%	51/59 86%	114/125 91%	28/34 82%	10/11 91%	59/68 87%
	227					

[a] NT = not tested.

chemicals. Complete agreement existed for 44 of these (85%). Divergent results are listed in Table 2. The poorest level of agreement between investigators is evident in the results of the alkyl halides. Differences were obtained for dieldrin, chlordane, and cyclophosphamide. Mitchell also detected DNA repair in cells exposed to folpet, yet no carcinogenicty has been demonstrated in animal studies.[28] Equally interesting was the agreement of three investigators on the lack of DNA damage or repair following exposure to DDT. Williams[34-36] also found DDE negative in the rat liver cell DNA repair system. Since the carcinogenicity of many of the alkyl halides resides primarily in the induction of "liver nodules", a debatable topic in itself, mechanistic approaches such as DNA damage and repair and mutagenesis may provide a clearer understanding of their biologic activity, i.e., toxin, promotor, or carcinogen.

B. In Vivo Systems

Much less attention has been given to the development and validation of in vivo DNA damage and repair assays. This is somewhat puzzling, when one realizes that proper metabolic activation represents an essential aspect of predictive short-term tests. In vivo systems utilize natural pathways of activation and detoxification. They also provide a mechanism for predicting target organs for carcinogens, as well as a tool for evaluating organ-specific toxicity for the presence of DNA damage.

While most reports of DNA damage following in vivo exposure have utilized ASG, there have been no publications concerning the use of this technique for screening a broad range of chemicals in a series of organs. Table 3 outlines the carcinogens, experimental systems, and results of most in vivo DNA damage studies that have employed gradients. These studies have usually evaluated one to four chemicals in one to four organs. Additional in vivo studies employing ASG will be needed before this technique can assume a screening role. Gradients also have the disadvantages of being time consuming, variable, and frequently having less than ideal recovery of radioactivity.

We have recently modified the alkaline elution method to detect carcinogen-induced DNA damage in various tissues following in vivo exposure.[24] The results are illustrated in Table 4. In general, a high degree of correlation existed between tissues with increased DNA elution following in vivo exposure to procarcinogens and target organs for tumor susceptability. This most likely denotes the metabolic competence of target organs to generate electrophils that bind to DNA. The correlation was much less evident for chemicals not requiring metabolic activation. Increased DNA elution was present in target and nontarget organs with compounds

TABLE 2

Divergent Results of DNA Damage and Repair Assays

Class and chemical	DNA repair (UDS)				Alkaline elution
	Mitchell[22,28]	Stich[25,29,30]	Williams[34-36]	Ahmed[1]	Swenberg[31,32]
Aromatic amines					
2-Aminobiphenyl	−	+	−	NT[a]	NT
Alkyl halides					
Dieldrin	NT	+	−	+	−
Chlordane	NT	+	−	−	NT
Cyclophosphamide	NT	+	NT	NT	−
Polycyclic aromatics					
Benzo(a)pyrene	+	−	+	NT	+
Benz(a)anthracene	+	+	−	NT	NT
Miscellaneous aliphatics and aromatics					
1′-Hydroxysafrole	NT	+	NT	NT	−
Metallic salts					
Lead acetate	−	+	NT	NT	NT

[a] NT = not tested.

TABLE 3

In Vivo DNA Damage and Repair — Alkaline Sucrose Gradients

Carcinogen	Dose range (mg/kg)	Route	Species	DNA damage	Comments	Ref.
4-NQO	40—80	s.c.	M	Lung > kidney > liver	4-AQO negative	19
	25—100	s.c., i.p.	R	Lung > liver > kidney		4
	100	oral	M	Gastric mucosa	6-Me NQO negative	18
DMN	1—10	i.p.	R	Liver	Only tissue, slow repair	6
	1—8	s.c.	M	Liver > kidney > lung		19
	15	oral	M	Liver	Negative stomach	18
2-AAF	11	oral	M	Liver	Negative stomach	18
N-Ac-AAF	13.75	oral	M	Liver, stomach	Equimolar	18
MAM	1—25	i.p.	R	Liver	Only tissue	6
	35	i.p.	R	Intestine	Only tissue	12
MNNG	Not given	oral	M	Stomach, Liver	Only tissue	18
MNUT	50—200	i.p.	R	Liver	Only tissue	6
MNU	20—160	i.p.	R	Liver	Only tissue	6
	0.1—0.5 mg	intravesicular	R	Bladder	Only tissue	5
ENU	20	i.p.	R	Brain	Only tissue	11
3'-Me-DAB	0.05%	diet	R	Liver	Only tissue CsCl gradients	9
3-2' Dimethyl-4-aminobiphenyl	Not given	s.c.	R	Intestine	Only tissue	12
HN$_2$	3	i.p.	R	Intestine	Only tissue	12
DMH	60	i.p.	R	Intestine	Only tissue	12
Mitomycin C	4	i.p.	R	Intestine	Only tissue	12

such as MNU and MMS. Not all carcinogens tested caused an increase in DNA elution from target organs. Nitrosohexamethyleneimine causes liver tumors, yet no increase in elution could be demonstrated, even at doses that caused liver necrosis. Likewise, chloroform treatment failed to increase the elution rate of DNA from rat kidney. Additional carcinogens and noncarcinogenic organ-specific toxins need to be evaluated with this method. Currently, however, it represents the best validated in vivo approach for detecting DNA damage and repair.

Two other systems involving DNA deserve mention. Sega et al.[26,27] have examined DNA repair in spermatids. Repair was detected following treatment of mice with MMS, EMS, PMS, MNU, hycanthone, cyclophosphamide, and ethylenamine. No repair was detected when animals were treated with MNNG, DMN, DEN, benzo(a)pyrene, or DMBA. This system provides a method for evaluating potential heritable damage. It would not be very useful for predicting carcinogenic activity in parental somatic cells. Through some elusive mechanism, treatment of mice with several procarcinogens resulted in inhibition of testicular DNA synthesis.[8] Friedman and Staub[8] demonstrated decreased DNA synthesis following treatment with MNU, DMH, DEN, MCA, Safrole, CdCl$_2$, 2-AAF, and dibutylnitrosamine. No decrease in DNA synthesis was seen when animals were treated with anthracene, diphenylnitrosamine, piperonyl butoxide, potassium cyanide, dinitrophenol, and lead acetate.

In summary, the mammalian DNA damage and repair assays appear to be good predictors of carcinogenic potential. While there are some

TABLE 4

In Vivo DNA Damage — Alkaline Elution[a]

Chemical	Dose Range (mg/kg)	Routes	DNA elution[b]								
			Liver	Lung	Kidney	Brain	Thymus	Duodenum	Stomach	Bone marrow	Mammary
Dimethylnitrosamine[c]	10—40	i.p.	+++	+	±	−	±	−			
Diethylnitrosamine[c]	40—80	i.p.	+++	−	±	−	−	+			
N-Nitrosohexamethyleneimine[c]	25—100	i.p., oral	−	−	−			−			
2-Acetylaminofluorene[d]	10—25	i.p.	+	−	−		−	−	−		
N-hydroxy-acetylaminofluorene[d]	20—40	i.p.	+	−	−		−				
Benzidine-HCl[d]	250	i.p.,s.c.	+	−							
Aflatoxin B$_1$[e]	0.5—1	i.p.	++	−	++		+			−	
1,2-Dimethylhydrazine[c]	20—100	s.c.	++	−	±		−	++			
Azoxymethane[c]	30—100	s.c.	+++					++			
4-Nitroquinoline-1-oxide[c]	25—100	s.c.	−	++	±	+					
	100	i.p.	±	+	±						
7,12-Dimethylbenz(a)anthracene[d]	75—200	Oral	−	±	−			−	++		±
Methylnitrosourea[f]	10—160	i.p.,i.v.	++	+	+++	++	++				++
Ethylnitrosourea[f]	50—100	i.p.,i.v.	+	+	±	+	−				++
Methyl methanesulfonate[c]	100—200	i.p.	+++	+++	+++	+++	+++	+++	+++		
Ethyl methanesulfonate[c]	100	i.p.	+		±		+				
N-methyl-N'-nitro-N-nitroso-guanidine[e]	250—500	Oral	++	−	−			++	++		
β-Propiolactone[d]	100—500	i.p., oral	++					+	+		

TABLE 4 (continued)

In Vivo DNA Damage — Alkaline Elution[a]

Chemical	Dose Range mg/kg	Routes	DNA elution[b]								
			Liver	Lung	Kidney	Brain	Thymus	Duodenum	Stomach	Bone marrow	Mammary
Azaserine[c]	50—250	i.p.	+++		++						—
Streptozocin[f]	50—100	i.v.	++	—	++	—					+
Cyclophosphamide[c]	5—80	i.p.	—								
Ethyl carbamate[c]	25—500	i.p.	—		±	+					
Chloroform[d]	200,400	Oral			—						

[a] Maximum increase in [³H]DNA elution is shown for each tissue examined. Exposure time ranged from 1 to 24 hr. All tissues studied did not exhibit DNA damage at each dose or exposure time investigated. Assay procedures were as described in "Materials and Methods".

[b] +++ = Increase elution > 50; ++ = increase elution > 20; + = increase elution > 7.5; ± = increase elution > 4.0; — = increase elution < 4.0.

[c] Administered in saline, controls received a similar volume of saline.

[d] Administered in DMSO, control received DMSO.

[e] Administered in corn oil, control received corn oil.

[f] Administered in 0.06 M sodium citrate-0.08 M k₂HPO₄, pH 4.2; control received vehicle.

From Petzold, G. L. and Swenberg, J. A., *Cancer Res.*, 38, 1589, 1978. With permission.

differences in results with the different assays, the overall correlation of test results is quite good. Methodology will continue to improve as a wider range of compounds are evaluated. Comparisons of results between assays would be greatly facilitated by the availability of a central clearinghouse. Assay results could be submitted as soon as they are complete, avoiding the delays associated with publishing. A major problem of such a system would be the potential for premature judgment of a compound's carcinogenicity. The DNA damage and repair assays, along with the other short-term tests, are not meant to replace well-run animal studies. The ultimate decision of whether or not a chemical is carcinogenic depends on the latter and proper epidemiological studies of human exposure. The short-term assays can be extremely useful in setting priorities for long-term animal tests and epidemiology, as well as providing early warnings of potentially dangerous chemicals.

REFERENCES

1. Ahmed, R. E., Hart, R. W., and Lewis, N. J., Pesticide induced DNA damage and its repair in cultured human cells, *Mutat. Res.*, 42, 161, 1977.
2. Casto, B. C., Pieczynski, W. J., Janosko, N., and Dipaolo, J. A., Significance of treatment interval and DNA repair in the enhancement of viral transformation by chemical carcinogens and mutagens, *Chem. Biol. Interact.*, 13, 105, 1976.
3. Cox, R., Damjanov, I., Abanoli, S. E., and Sarma, D. S. R., A method for measuring DNA damage and repair in the liver *in vivo*, *Cancer Res.*, 33, 2114, 1973.
4. Cox, R. and Irving, C. C., Damage and repair of DNA in various tissues of the rat induced by 4-nitroquinoline-1-oxide, *Cancer Res.*, 35, 1858, 1975.
5. Cox, R. and Irving, C. C., Effect of N-methyl-N-nitrosourea on the DNA of rat bladder epithelium, *Cancer Res.*, 36, 4114, 1976.
6. Damjanov, I., Cox, R., Sarma, D. S. R., and Farber, E., Patterns of damage and repair of liver DNA induced by carcinogenic methylating agents *in vivo*, *Cancer Res.*, 33, 2122, 1973.
7. Frei, J. V., Some mechanisms operative in carcinogenesis — a review, *Chem. Biol. Interact.*, 13, 1, 1976.
8. Friedman, M. A. and Staub, J., Inhibition of mouse testicular DNA synthesis by mutagens and carcinogens as a potential simple mammalian assay for mutagenesis, *Mutat. Res.*, 37, 67, 1976.
9. Goodman, J. I., Hepatic DNA repair synthesis in rats fed 3'-methyl-4-dimethylaminoazobenzene, *Biochem. Biophys. Res. Commun.*, 56, 52, 1974.
10. Goth, R. and Rajewsky, M. F., Persistence of O^6-ethylguanine in rat-brain DNA: correlation with nervous system-specific carcinogenesis by ethylnitrosourea, *Proc. Natl. Acad. Sci. U.S.A.*, 71, 639, 1974.
11. Hadjiolov, D. and Venkov, L., Strand breakage in rat brain DNA and its repair induced by ethylnitrosourea *In Vivo*, *Z. Krebsforsch.*, 84, 223, 1975.
12. Kanagalingan, K. and Balis, M. E., *In vivo* repair of rat intestinal DNA damage by alkylating agents, *Cancer*, 36, 2364, 1975.
13. Kirtikar, D. M. and Goldthwait, D. A., The enzymic release of O^6-methylguanine and 3-methyladenine from DNA reacted with the carcinogen N-methyl-N-nitrosourea, *Proc. Natl. Acad. Sci. U.S.A.*, 71, 2022, 1974.
14. Kleihues, P., and Bucheler, J., Long-term persistence of O^6-methylguanine in rat brain DNA, *Nature (London)*, 269, 265, 1977.
15. Kleihues, P., Lantos, P. L., and Magee, P. N., Chemical carcinogenesis in the nervous system, *Int. Rev. Exp. Pathol.*, 15, 153, 1976.
16. Kohn, K. W., Erickson, L. C., Ewig, R. A. G., and Friedman, C. A., Fractionation of DNA from mammalian cells by alkaline elution, *Biochemistry*, 15, 4629, 1976.
17. Kohn, K. W., Friedman, C. A., Ewig, R. A. G., and Iqbal, Z. M., DNA chain growth during replication of asynchronous L-1210 cells. Alkaline elution of large DNA segments from cells lysed on filters, *Biochemistry*, 13, 4134, 1974.
18. Koropatnick, D. J. and Stich, H. F., DNA fragmentation in mouse gastric epithelial cells by precarcinogens, ultimate carcinogens and nitrosation products: an indicator for the determination of organotrophy and metabolic activation, *Int. J. Cancer*, 17, 765, 1976.
19. Laishes, B. A., Koropatnick, D. J., and Stich, H. F., Organ-specific DNA damage induced in mice by the organotropic carcinogens 4-nitroquinoline-1-oxide and dimethylnitrosamine, *Proc. Soc. Exp. Biol. Med.*, 149, 978, 1975.
20. Lindahl, T., New class of enzymes acting on damaged DNA, *Nature (London)*, 259, 64, 1976.
21. Miller, J. A., Carcinogenesis by chemicals: an overview, *Cancer Res.*, 30, 559, 1970.

22. Mitchell, A. D., personal communication.
23. Nicoll, J. N., Swann, P. F., and Pegg, A. E., Effect of dimethylnitrosamine on persistence of methylated guanines in rat liver and kidney DNA, *Nature (London)*, 254, 261, 1975.
24. Petzold, G. L. and Swenberg, J. A., Detection of DNA damage induced *in vivo* following exposure of rats to carcinogens, *Cancer Res.*, 38, 1589, 1978.
25. San, R. H. C. and Stich, H. F., DNA repair synthesis of cultured human cells as a rapid bioassay for chemical carcinogens, *Int. J. Cancer*, 16, 284, 1975.
26. Sega, G. A., Owens, J. G., and Cumming, R. B., Studies on DNA repair in early spermatid stages of male mice after *in vivo* treatment with methyl-, ethyl-, propyl-, and isopropyl methanesulfonate, *Mutat. Res.*, 36, 193, 1976.
27. Sega, G. A., personal communication.
28. Simmon, V. F., Mitchell, A. D., and Jorgenson, T. A., Evaluation of Selected Pesticides as Chemical Mutagens — *In Vitro* and *In Vivo* Studies, Environmental Health Effects Research Series EPS-600/1-77-028, May 1977.
29. Stich, H. F., Kieser, D., Laishes, B. A., San, R. H. C., and Warren, P., DNA repair of human cells as a relevant, rapid and economic assay for environmental carcinogens, *Gann Monogr. Cancer Res.*, 17, 3, 1975.
30. Stich, H. F., personal communication.
31. Swenberg, J. A., Petzold, G. L., and Harbach, P. R., *In vitro* DNA damage/alkaline elution assay for predicting carcinogenic potential, *Biochem. Biophys. Res. Comm.*, 72, 732, 1976.
32. Swenberg, J. A., unpublished data.
33. Verly, W. G., Maintenance of DNA and repair of sites without base, *Biomedicine*, 22, 324, 1975.
34. Williams, G. M., Carcinogen-induced DNA repair in primary rat liver cell cultures; a possible screen for chemical carcinogens, *Cancer Lett.*, 1, 231, 1976.
35. Williams, G. M., Detection of chemical carcinogens by unscheduled DNA synthesis in rat liver primary cell cultures, *Cancer Res.*, 37, 1845, 1977.
36. Williams, G. M., personal communication.
37. Williams, J. R., Little, J. B., and Shipley, W. U., Association of mammalian cell death with a specific endonucleolytic degradation of DNA, *Nature (London)*, 252, 754, 1974.
38. Zubroff, J. and Sarma, D. S. R., A nonradioactive method for measuring DNA damage and its repair in nonproliferating tissues, *Anal. Biochem.*, 70, 387, 1976.

Section II
Practical Guidelines for the Use and Interpretation of Short-term Testing for Mutagens/Carcinogens

RECOMMENDATIONS FOR PRACTICAL STRATEGIES FOR SHORT-TERM TESTING FOR MUTAGENS/CARCINOGENS

B. E. Butterworth

TABLE OF CONTENTS

I. Introduction 89

II. The Value of Short-Term Tests 89

III. Which Tests Should Be Performed? 90
    A. Battery vs. Tier System 90
        1. Discussions by Eric Longstaff, Donald Clive, and Roman J. Pienta 90
    B. Resources and Expertise Necessary for Running the Tests 91
        1. Discussions by Robert Kuna, Marvin S. Legator, William G. Thilly, Verne A. Ray, Donald Clive, and Roman J. Pienta 92
    C. Degree of Development of the Test 95
        1. Discussions by Marvin S. Legator and William Sheridan 96
        2. A Comparison of In Vivo and In Vitro Metabolic Activation Systems — Gary M. Williams 96
            a. Metabolic Activation In Vitro 96
            b. Correlation of In Vitro Results with In Vivo Carcinogenicity 97
    D. Is the Test a Valid Test? 97
        1. Discussion by Marvin S. Legator 98
        2. Status of NCI Validation Studies — Virginia Dunkel 98
        3. The Standardization and Validation Process — Verne A. Ray 98
        4. The Question of Mutagenicity Per Se 99

IV. Factors to Be Considered When Evaluating the Results of the Tests 99

V. Possible Courses of Action for a Positive Response 100

VI. Possible Courses of Action for a Negative Response 101

References 101

I. INTRODUCTION

These guidelines were developed by the Chemical Industry Institute of Toxicology prior to the conference. They were debated and in many cases modified by suggestions from the Workshop participants. There was considerable discussion regarding several of the main points. In these instances, a summary or transcript of that discussion is presented following those points.

II. THE VALUE OF SHORT-TERM TESTS

Traditional animal exposure tests for carcinogenicity require at least 2 years and can cost in excess of $300,000 for each chemical tested.

Even then, the results of such tests may be difficult to relate to human health hazard. Other information that would bear on the mutagenic/carcinogenic potential of a chemical would be of tremendous value in defining a possible problem or in supplementing the results of long-term tests. The so-called "short-term" tests of mutagenicity/carcinogenicity may serve this purpose. However, in arriving at an assessment of *health hazard,* additional information such as the potency of the compound and the nature and extent of exposure to it are essential.

Some uses of the short-term tests include:

1. Results from the tests can be of value in setting priorities for the selection of chemicals to be subjected to long-term studies. If there is concern that one of a dozen chemicals in the workplace might be a carcinogen, they can all be run through the short-term tests and more emphasis placed on those that are positive.
2. The tests may be used as tools to "design out" carcinogenicity from products in development. All things being equal, the putative nonmutagenic analogues of new products would be more attractive for development. Several people mentioned that designing out toxicity is difficult. Often a decrease in toxicity results in a concomitant decrease in the desired biological effect. Nevertheless, the short-term tests do provide a rapid feedback to the synthetic chemist and the test results should be considered as part of the whole picture.
3. Possible mutagenic impurities in a sample can be analyzed. The short-term nature of the tests allows all samples from a factionated preparation to be analyzed for mutagenic activity. In some cases the offending chemical is only an impurity and exposure can be eliminated by using a higher grade material, a different route of synthesis, or other modifications.
4. Care should be exercised to use the tests only in appropriate circumstances. For example, in monitoring air samples in the workplace where the biologically characterized pollutants are known, it is much more straightforward to use a chemical detector rather than to gather samples, elute them, and try to run a quantitative *Salmonella*/microsome assay (Ames test) on each.[1]

III. WHICH TESTS SHOULD BE PERFORMED?

A. Battery vs. Tier System

There are two approaches to short-term testing. One is the tier system in which chemicals are screened with an initial, easily performed, inexpensive test. Those chemicals that turn up as positive are then subjected to progressively more lengthy and expensive tests. The second approach is the battery system, in which the compound is more or less simultaneously subjected to several tests. The battery approach is superior. All tests yield a fraction of false positive and false negative results, and no individual test is reliable enough to stand alone. For example, the potent carcinogen hexamethylphosphoramide (HMPA) is negative in the Ames test but is detected in mammalian cell systems. If the tier system had been used with the Ames test as the first tier, then this potent carcinogen would have gone undetected through the system. It is recommended that two or three of the following types of tests be used as a battery on all compounds examined: bacterial mutagenesis, cultured mammalian cell mutagenesis, cell transformation, DNA damage, and whole animal mutagenesis or chromosomal damage. Many such tests are described in detail in Section I. Each testing program must, of course, be independently formulated to meet specific needs. General principles of selection would be

1. Choose a variety of types of tests.
2. Choose at least one system that deals with mammals or mammalian cells.
3. Keep in mind that preliminary results suggest that cell transformation systems will be especially accurate predictors of the carcinogenic potential of chemicals.

1. Discussions by Eric Longstaff, Donald Clive, and Roman J. Pienta

On the whole, the consensus was that the battery approach was superior and that results from several valid short-term tests gave a more reliable picture than any single test. Several people presented data showing that various tests differed with respect to the classes of carcinogens for which they were sensitive.

Dr. Longstaff listed several carcinogens that were negative in the Ames test: saccharin, *o*-to-

TABLE 1

False Negative Responses

Hamster cell transformation without activation	Bacterial mutagenesis with activation
4-Aminoazobenzene	
3-Methoxy-4-aminoazobenzene[a]	
2-Nitrofluorene[a]	
Diethylnitrosamine[a]	
p-Rosaniline[a]	p-Rosaniline
Natulan[a]	Natulan
Urethan[a]	Urethan
Auramine[a]	Auramine
	Acetamide
	Thiourea
	1,2-Dimethylhydrazine
	Ethionine
	Succinic anhydride
	Safrole
	3-Amino-1,2,4-triazole
	Bis(p-dimethylamina)-diphenylmethane
	Lead acetate
	Beryllium sulfate
	Titanocene dichloride
	Nickel sulfate hexahydrate

[a]Positive with activation.

luidine, safrole, auramine, HMPA, diethylstilbestrol, 3-aminotriazole, hydrazine, chloroform, and urethane.

Dr. Clive presented his preliminary data indicating that several carcinogens were detected in the L5178Y mouse lymphoma mutagenicity assay, which were missed in the Ames test: diethylstilbestrol, natulan, mitomycin C, DDE (2,2-bis(p-chlorophenyl)-1,1-dichloroethene), HMPA, and saccharin.[2]

Similarly, Dr. Pienta presented data showing that the spectrum of false negative responses in the Golden Syrian hamster embryo cell transformation assay was quite different from the false negative responses in the Ames test[3] (see Table 1).

It is clear from these examples that reliance on any single test as a first tier would result in selectively missing those classes of chemicals for which that particular test was not sensitive. Routine testing will include chemicals from all kinds of classes about which we have no carcinogenic information. A screen should, therefore, utilize several different tests.

Dr. Longstaff presented summary data from the validation studies in which Imperial Chemical Industries, Ltd. measured the performance of various short-term tests with 120 compounds of known carcinogenicity or noncarcinogenicity:[4]

	Carcinogens positive in one test	Noncarcinogens negative in all tests
	(%)	(%)
Cell transformation	91	97
Ames test	91	94
Cell transformation and Ames test	99	91

The combination of Ames plus cell transformation was superior than either single test in detecting the known carcinogen (91 vs. 99%). Furthermore, in those cases where there was a positive in *both* tests, the compound *always* turned out to be carcinogenic — there were no false positives.

The disadvantage of using multiple tests is that the chance of picking up a false positive in one of the tests will increase with the number of tests used. This is illustrated in the data shown above and was presented in theoretical statistical arguments by Dr. Longstaff and Dr. Cook. Nevertheless, the advantages of the battery outweigh this disadvantage, and a rational examination of the whole picture should minimize this problem.

Of course, until such time as a number of reliable tests are available, those tests that we do have can be profitably used, even if they are limited to one or two. Several people made the point that under certain circumstances, the results from a single test could stand alone. Dr. Longstaff pointed out that often testing is not random and testing is recommended because of a structural similarity between the new chemical and a known mutagen/carcinogen. In this case, if a given test recognizes the parent carcinogen as positive, the results on the new test chemical will be meaningful.

B. Resources and Expertise Necessary for Running the Tests

A major concern of many in industry was that the battery approach was a good idea —

however, the number of validated tests and qualified laboratories to perform them was severely limited. In addition, the battery approach means substantially increased costs. These points are well taken. Even the Ames test is complicated and spurious results are easily obtained if attention is not paid to detail. Getting the test set up requires months and a substantial investment in facilities and personnel. The cell culture assays are even more complex and the supply of experienced personnel is limited. Some of the cell transformation assays can only be performed by a handful of laboratories across the country. These concerns were illustrated by the following remarks.

1. Discussion by Robert Kuna, Marvin S. Legator, William G. Thilly, Verne A. Ray, Donald Clive, and Roman J. Pienta*

Butterworth — The consensus seems to be that a battery of reliable tests is a good way to approach the problem. However, I feel that the resources in the country to do such a battery of tests are very limited and it can be frustrating for an organization to find the people and physical facilities to do the kind of thing that we're suggesting, especially if you are trying to set up your own in-house program.

Kuna — There are quite a few companies who are now doing some kind of mutagenic testing. One philosophy is to get quick results on many product candidates at a very early stage in development. An Ames test is being used quite frequently for this. At Exxon, they prefer not to run this way. They prefer to technically get one compound that has all the activities they want to see in the end use, then send it down to us and ask for a toxicological evaluation. They want procedures run on a toxicological evaluation which are acceptable and meaningful for the governmental agencies and have acceptable scientific backing for the data that they have. For this reason, at this time at Exxon we have not felt that the Ames test is appropriate for this type of compound coming down stream. We have, I think, come to the conclusion that a battery of tests will probably be the best road to go for mutagenic testing. Byron (Butterworth) has asked, what type of battery should be used? The meeting is farily well-broken up into people who are responsible for this testing on the industrial side and people who are actually formulating the methodology and providing us with the ammunition to do the testing. The thing that impresses me right now is that if the group of the doers here had to sit down and take five tests that they would use across-the-board for a battery of testing, from what I heard yesterday, I do not think there are five tests at this time that everybody would agree have been fully evaluated. This means that on the industial side, we are not testing our chemicals, we are testing the method — which is really not the thing we want to do.

At Exxon again, we do not have in-house capabilities, as many of the other chemical companies represented here have. This means we have to do all our work outside. Going outside, I look at the testing laboratories, I look at the expertise which is in the testing laboratories, and I ask myself, if we had a battery of five tests that were regulated by government or self-regulated, is there enough expertise outside to take the load and do the testing adequately? I have reservations that there is that type of expertise. Exxon belongs to a series of trade organizations. I am the Exxon representative for one of these. Through this organization this year we have put out two different compounds for cytogenetic testing. One compound went to one contract laboratory, another compound went to three different contract laboratories. For the compound that went to the three different testing laboratories, we got three different answers, and we're left with that. For the other compound, the report is now being rewritten because it was unacceptable at the time we received it. We're in an area of science where I think we should heed some of what Bern (Schwetz) said, in that we should make sure we know where we're going before we make any big steps.

Participant — You seem to address yourself to two issues; one concerns which tests to use in a battery and what these tests mean, and the other is the availability of competent labs to run them. I should like to address myself to the issue of some of the "accepted" toxicity

---

* The transcript presented here is an edited version of the original. Some sections have been omitted and wording is not necessarily verbatim.

tests. When did you start to use 3-month toxicity studies? How much validation was needed before you would accept a 3-month toxicity study in a rat and a dog? In terms of long-term bioassay, it came out this morning that a 2-year study that was done 5 to 10 years ago may not have the same type of detail that would be in a present-day protocol. So here is a test that many people rely on and yet has been in a dynamic state. It's changed as science has changed and as technology has advanced, so I believe that many of the mutagen tests that are being proposed or suggested may not be fully validated, but yet they are being used and will change, and the validation procedures that are being used in the present day may not be the same ones that will be 5 years from now.

Kuna — I think if you got a group of scientists together as in this meeting, and you presented a 90-day study, you would have a little more agreement than we had here yesterday. With most of the procedures that were presented, it was admitted that more work had to be done to back them up. I think the mutagenesis testing has taken tremendous strides in the last 2 years, but I think there's much more that has to be done.

Legator — If we had a group of teratologists together, they would have an even more difficult time in arriving at a consensus for protocols. I think we have something here to offer that you probably don't have in any other area of toxicology — we have the ability to find out what's happening on a molecular level, which you can't do in almost any other area of toxicology. I don't really think that we would have a difficult time in outlining the kinds of tests that one should do within a battery of tests. I have felt that there are excellent contract laboratories that do a good job of testing and the cost, as I understand for a whole battery of tests, which includes dominant lethal, cytogenetics and a number of in vitro tests, comes to about $12,000 to $15,000, which is rather nominal compared to what you're putting into any other area of toxicology. I think we have something here that is a challenge to other areas of toxicology.

Brusick — If you look at various types of mutagenicity assays, such as cytogenetics, dominant lethal, Drosophila and Ames test, the number of chemicals that have been tested probably exceeds the number tested for validation in most of these other types of toxicological tests, such as teratology. The problems come, I still think, in terms of interpreting the data. There are many compounds that have been tested multiple times in 2- or 3-year feeding studies and have produced different results. Even for LD_{50} determinations, you can send out one compound to several different laboratories for an acute LD_{50}, and I guarantee you that the results will not be the same value. It is important to know the magnitude of the difference in the responses, and whether those differences can be explained by the different approaches taken by laboratories for evaluating that response. When we address the problem of standardization and defining acceptable levels of activity, I think we can eliminate some of the problem.

Swenberg — You've got to keep your options open so that depending on what the need is, you can use what appears to be an appropriate number of tests. In some cases, you may not want to have a large battery of tests, you can't afford it, you don't have the time and the manpower.

Thilly — May I continue on the point of manpower and ask just how many people in this country are competent at running any of the various cell transformation tests? How many people are competent to go into the laboratory and run the cell mutation tests, and how many people in a contract laboratory are competent to analyze DNA repair tests, particularly of the DNA elution type, in addition to the unscheduled DNA synthesis, and what kind of compound load can be handled by the U.S. to test compounds in this coming year?

Participant — Well, I don't think there's any doubt that what you're saying is very true, but most of us find ourselves in a situation that is not of our creating; we have very little choice.

Thilly — I want to ask the question so that we can put it into persepective here at the end of the meeting before we run off and say yes, use a battery of tests. Let's take my own human lymphoblast test. There's only about six or seven people who have ever done that in their entire lives. Maybe Abe (Hsie) has got twice that number who have worked in his lab in Oak Ridge and can do the CHO assay with the competence that was presented yesterday. Don (Clive) can tell us how many people he has trained that he feels are competent to run his mouse lymphoma assay. How many compounds can be run in a reasonable battery?

Legator — I would like to ask Dave Brusick what he does as a contract lab. I think you're perfectly right; we all come up with a system that we work out in our laboratory and we're the only ones who can do it. On the other hand, I think that there is sufficient expertise in contract laboratories and in in-house capabilities to do much of the work.

Brusick — We are in a situation where there is a need and the need has not necessarily developed all in one sector. The need comes from both the need to answer scientific questions and the need to meet either the explicit or implicit demands of regulatory agencies, and in some cases these don't necessarily coincide. There has been a demand from clients to provide tests such as the mouse lymphoma assay, which Don (Clive) admits requires more research and may not be ready for routine use. We provide this service to the best of our ability and explain the results as realistically as possible. We consult with individuals like Don (Clive) and other people, and we realize that it may not be absolutely defined as a system that will pick up all types of chemicals. We also accept the fact that the protocol we're using might not be the final protocol one could use.

Our laboratory now consists of approximately 40 people. We do have some people who have not been involved in this area for their total scientific career. It is difficult to find people who have been experienced in mutagenicity testing. It's not as simple as getting a geneticist or cell biologist or someone of that type and saying: "here, we are going to test chemicals." You need someone who has a feel for chemical mutagenesis and what chemicals can do in cells, animals, and bacteria. So, again, there is a problem of just getting the right kind of people.

I can identify for you the array of tests we offer: a complete series of microbial tests and mammalian cell culture tests, such as the mouse lymphoma system. We are trying to develop and incorporate into cultured cell tests an assay for measuring sister chromatid exchange and chromosome aberrations simultaneously, so that one can dose cells and detect three different types of endpoints from the same set of dosed cells. This makes a stronger assay and lends more confidence to an indication of an effect. We do unscheduled DNA synthesis in WI38 cells, and on a call basis, transformation in BALB/3T3 cells. We attempt to relate to the client that some of these assays are not routine assays. We conduct the tests and we're trying to improve the assays in terms of a more routine assay for the contract laboratory. If there is a desire for new tests, they can be conducted under established protocols on an indefinite time period situation. From there we can go into animal models such as dominant lethal, in vivo cytogenetics, the micronucleus test, and heritable translocation. We are involved in developmental tests such as the recessive spot test and the Drosophila, sex-linked, recessive lethal. There are other contract laboratories that offer this type of array. This capability may not be sufficient, however, for all interested clients, if such tests had to be initiated for a number of companies all at the same time. I think there would be some problem in developing sufficient resources nationwide to handle the whole testing load if all of these assays were required in the very near future.

Pienta — We always extend an invitation to anyone who would like to learn our cell transformation assay to come and see it. We show everyone all the problems and all the morphological criteria we use. We show them all the tricks, all the nuances, and I know there are several people in the audience who have taken me up on that offer.

Ray — What would your estimation be of the time for someone who has tissue culture capabilities to learn to set it up?

Pienta — We can teach someone to look at all the parameters of the test in a week or so. I estimate it would probably take 3 to 6 months for a lab to set it up. There are two basic things you have to do: you have to first identify the number of primary cultures that react, and once you do that you have to develop a stock that's useful. People who have been trying to work this assay the last 2 years have had a lot of trouble with it because some of the animal studies weren't the best. We have 250 ampules of this particular batch of cells that should last us for a couple of years.

Ray — Would it be fair to say that it would take an industrial laboratory, even with good competent people and having that training, a year's worth of activity before they could reach a competence to invest their own compounds in that assay?

Pienta — Six months to a year. You have to realize that it is expensive, you need exhaustable laminar flow hoods, you need a good air balance, and good cell culturists who appreciate that these things are alive. You can't put them on a shelf on Friday night and expect to see them there on Monday; they might have to be looked at on Sunday. We do our tests blind, every one of our compounds are tested coded and we do pretty well. I still don't like to say correlations of 90 or 80%, because I think it is too great a generalization. We've been able to isolate about 25 of these colonies, every one of which has criteria that are generally accepted as in vitro transformants; they grow in soft agar and every one of them induces tumors in random-bred hamsters that have been immune suppressed by X-ray or steroids. So we are quite confident. Our numbers appear to be kind of low when we look at the frequencies. People here are accustomed to looking at hundreds of mutants but a plate containing 10^8 bacteria is really not the same as one containing 500 target cells of heterogenous population, perhaps maybe only 10% of which can be transformed under any circumstance. When we see one or two or several colonies, we're confident they are real. It's not easy, but once you get it to work, it works.

Clive — I'd like to add a little bit to that. Roman (Pienta) painted the proper picture. These are short-term tests and that is not an excuse for doing sloppy work. I think too many people are taking for granted, because they are short-term, that anybody can jump in with a minimum amount of effort and expect it to work immediately and start applying it. This cannot be done. The best experience I've had with people coming to my lab, as far as getting it to work, is by somebody who has admitted to himself and to others that he has absolutely no ability and follows every procedure to the letter.

Pienta — We have outlined our protocols step by step from A to Z. If you follow those religiously, it works. We had someone visit us a couple of weeks ago and we said, plate the cells and you can't use them if they have to take more than 3 days to become 90% confluent. If they don't look as good as you think they should, throw them out. I get a call from this person on Friday who says I put the cells down yesterday, they look like they are 80% confluent today, what am I going to do with them, I think I'll put them on minimal medium and let them hang over till Monday. Now, if he can't do the test, then he says, well we followed the procedure and it didn't work.

C. Degree of Development of the Test

Data generated by most basic research have little immediate effect on the every-day workings of business or people's lives. Data from tests that claim to predict carcinogenicity have an implication to human health and must be acted upon. Consequently, a great deal of thought should go into the validity and degree of development of the test that is chosen. It is recommended that tests, the results of which will be used in the decision-making process, *not* be run unless the significance of the results will be understood and one is prepared to be held to those results.

For example, the measurement of mutagenicity of human urine specimens has been proposed as a means of monitoring exposure to mutagens in the environment. However, several important questions remain to be answered:

1. The normal variations in the human population have not been established. The modifying effects of genetic differences, age, sex, diet, medication, smoking, use of hair dyes, etc. all have yet to be determined.
2. The amount of data reported for these systems is limited. Only a few chemicals or their metabolites have been tested in the urine assay. The spectrum of compounds that would show up in these tests has not been determined.
3. In experiments done to develop this test, large doses of carcinogen were administered by direct routes. Almost no data have been gathered on the degree of exposure necessary to elicit a response, nor the dependence on the time that elapses between exposure to the chemical and collection and analysis of the urine.
4. A comprehensive set of procedures for fractionating the urine and releasing conjugates has not been developed.

It would be inappropriate to administer such a test in the workplace until the above parameters were defined. The data generated would only add to the burden of unsettling, uninterpretable information that now exists.

1. Discussions by Marvin S. Legator and William Sheridan

Dr. Legator has had more experience than most with testing for mutagens in urine and feels the technique is extremely valuable because it begins to bridge the gap between in vitro systems and human models. He felt the technique was useful and had had few problems with false positives in their experience of testing hundreds of samples, primarily from industrial workers at Dow Chemical. Current work is beginning to shed more light on this subject.[5-8] The main point was, of course, not that the human urine test should never be used, but that it should not be used in the decision-making process until the results could be meaningfully interpreted.

There was considerable concern and agreement that much more research remains to be done with the activation portion of the short-term testing systems. Dr. Legator outlined some serious limitations to the Ames test because of questions about the S9 portion of the test:

1. Inability to standardize in vitro activation systems.
   a. Insufficient understanding of the relationship of enzymes to mutagenicity.
   b. Variability with the amount of S9 fraction used.
   c. Variability with chemical inducer of liver microsomal enzymes, administered beforehand to the animal.
2. Inability of in vitro systems to detect promutagens activated by means other than by liver microsomes.

Dr. Sheridan emphasized some of the pitfalls in extrapolating from the simple short-term tests to whole animals.

Sheridan — I'd like to make a comment on the whole question of activation. It seems to me that people have been taking a one-way street, that is they are using liver microsomes. It's a well known phenomenon, as any pharmacologist will tell you, that a substance may go to the liver and be toxified or detoxified and go a step further to the next organ and the reverse procedure will happen. So, where you do find a lack of correlation between your test-tube activation and the in vivo test in the whole animal, it may be the consequence of this kind of toxification/detoxification, again taking place at several different steps throughout the entire organism. In the whole picture of the Ames test and other tests where activation is required, people seem to ignore this point. I don't know whether it's intentional or whether they just haven't thought of it, but I think that there is good pharmacological evidence for this kind of thing happening in the animal and it should be taken into consideration here.

2. A Comparison of In Vivo and In Vitro Metabolic Activation Systems — Gary M. Williams

a. Metabolic Activation In Vitro

Perhaps the most critical feature of any in vitro screening system for chemical carcinogens is the competence for metabolic activation of procarcinogens. Rat liver possesses the most diverse capability for metabolic activation[9] and appears to be capable of metabolizing every procarcinogen, except those requiring modification by bacterial enzyme systems, such as cycasin. For this reason, rat liver microsomes have been the most useful supplement to in vitro systems for providing metabolic activation. However, it is to be expected that differences will exist between the metabolism performed by microsomes and that which occurs in the intact organ. One reason for this is that microsomal preparations, unless specifically modified, do not perform conjugation reactions and may be deficient in other detoxification pathways. Such deficiencies of microsomes may explain why pretreatment with microsomal-enzyme inducers enhances activation in vitro, but inhibits carcinogenesis in vivo; that is, the detoxification pathways induced by pretreatment, which result in decreased carcinogenicity, may be selectively lost in microsomal preparations, leaving a net enhancement of activation. Therefore, although microsomal mediation is of unquestionable value in establishing whether metabolism of a chemical can give rise to a mutagen, this metabolism may not accurately reflect the real in vivo situation. This complication makes species and tissue comparisons between microsomal-mediated activation and carcinogenesis subject to error and, indeed, there are now a number of examples of microsomal-mediated activation by microsomes from organs that are not the target for the carcinogens found to be activated.

b. Correlation of In Vitro Results with In Vivo Carcinogenicity

From the foregoing discussion, it may be suggested that microsomal-mediated assays may give rise to "false positive" results as a consequence of their deficiency in normal detoxification reactions. On the other hand, they also give "false negative" results for a number of reasons, including lack of appropriate metabolism and failure of the activated metabolite to reach the target macromolecule. However, in addition to "false negatives", it is also very likely that there will be *true negatives,* i.e., chemicals that produce tumors in vivo but are not mutagens or do not interact with DNA and, thus, are negative in the in vitro systems. Examples of such chemicals might be phenobarbital or DDT, which are promoters for rat liver tumor development and which increase the incidence of mouse liver tumors in strains with a spontaneous incidence of liver tumors. Consideration of the molecular structure of these chemicals does not reveal a likely electrophilic metabolite that could interact with cellular macromolecules in the manner now clearly understood for most chemical carcinogens. Therefore, I have suggested[10] that the operational definitions of carcinogens (i.e., agents that produce an increase in neoplasms in animals) should be reexamined with the object of developing mechanistic definitions. With this approach, carcinogens that generate electrophilic reactants that interact with DNA and other macromolecules may be classified as genotoxic carcinogens. Those operating by other mechanisms may provisionally be classified as epigenetic carcinogens or promoters until their mechanism of action is elucidated. In vitro systems will obviously be critical in identifying genotoxic carcinogens. However, as noted earlier, there are "false negatives" in in vitro systems and, therefore, means must be developed for distinguishing between these and true negatives that are not genotoxic carcinogens. To a large extent, structure will be a major criterion, but, in addition, the measurement of induction of DNA repair in primary cultures of heptocytes[11] may be useful in this regard. The reason for this is that, in our studies thus far, hepatocyte cultures have retained a broad representative metabolic capability characteristic of liver; because metabolism occurs in the target cell, as in vivo, this system may not suffer from the same limitations as those systems requiring microsomal mediation. Indeed, thus far, every hepatocarcinogen that has been tested is positive with the exception of phenobartibal. Therefore, I have suggested that phenobarbital and possibly the organochlorine pesticides are not genotoxic carcinogens.[10] Apart from the mechanisitic implications, the significance of such a concept, if it ultimately proves to be valid, is that the human hazards from such agents may be of quite a different nature than those from genotoxic carcinogens.

D. Is the Test a Valid Test?

Short-term tests are of value only insofar as they predict a hazard to human health and the environment. In the area of chemical carcinogenesis, animal models are chosen because results from such studies appear to be currently the best criterion for assessing potential hazard in man. The Ames test measures mutagenicity in selected strains of *Salmonella typhimurium.* It does not measure carcinogenicity. The value of a test such as the *Salmonella/*microsome assay lies in the demonstrated correlation with long-term animal studies. Many short-term tests have been proposed. All are in the developmental stages, and few have undergone correlation studies with large numbers of compounds. For example, Purchase and co-workers evaluated six short-term tests that purported to predict the carcinogenic activity of chemicals.[4] They reported that only two of the six tests showed a reasonable correlation with animal studies. It certainly would not make sense to reach decisions based on one of the four noncorrelating tests. All proposed short-term tests must be evaluated in order to define those chemical classes for which they are valid. This should be done with a recognized set of compounds comprising a carcinogen, a weak carcinogen, and a noncarcinogen in the same chemical class, such as *N*-nitrosodimethylamine/N-nitrosodi-n-butylamine/N-nitrosodiphenylamine for the nitrosamines and benzo(a)pyrene/benzo(e)pyrene/pyrene for the polycyclic aromatic hydrocarbons. The Ames test responds well to the polycyclic aromatic hydrocarbon carcinogens, but does not respond to the metal carcinogens. If one is testing a compound in a particular chemical class, whenever possible, tests should be chosen that respond correctly to that chemical class. However, many compounds do not fit into previously established classes. In these cases short-term tests should be chosen which have been shown to have a broad spectrum of response.

1. Discussion by Marvin S. Legator

Dr. Legator was critical of generalized statements that a system was "validated":

Legator — In the literature, although we often refer to the 90% figure, in terms of correlation between the *Salmonella* system and known carcinogens, the truth of the matter is that one can find in the literature correlations ranging from 44 to 90 + %. In our own analysis of these studies, we can, in part, attribute the variation in correlation either to the number of classes or number of compounds studied. The high correlations reported in the literature are probably not valid for one or more of the following reasons:

1. Failure to establish uniform and meaningful criteria for what is a carcinogen.
2. Failure to establish meaningful criteria for the response in the *Salmonella* system.
3. Failure to code samples.
4. Using literature studies to derive correlations.

2. Status of NCI Validation Studies

While it is anticipated that many of the proposed assays will become useful short-term tests, it must be remembered that substantial work remains to define systems and correlate results for most of the assays. Dr. Dunkel described the correlation studies that are in progress by the NCI.

Dunkel — The in vitro carcinogenesis program is currently both developing and validating a series of microbial and mammalian cell systems. The microbial assays include (1) mutagenesis in *Salmonella typhimurium* strains and *Escherichia coli* WP2, and (2) DNA repair in *E. coli* and pol A+ and pol A− strains. The mammalian cell assays include (1) mutagenesis in the L5178Y mouse lymphoma cell line at the thymidine kinase locus, (2) DNA repair in primary rat liver hepatocytes, and (3) transformation in systems using BALB/c 3T3 mouse cells, early passage Fischer rat embryo cells, Fischer rat embryo cells infected with Rauscher leukemia virus, and hamster embryo cells either treated directly in cell culture or exposed to the chemical transplacentally and then placed in culture and liver and skin epithelial cells. As now structured approximately, 90 compounds will be tested blind in all assays.

Clive — How long will it be before we know the results of the study?

Dunkel — It may be 2 years before we know the results of the blind studies.

3. The Standardization and Validation Process — Verne A. Ray

Dr. Ray expressed the concern of many that some tests could be brought to regulatory status before they were sufficiently characterized.

Ray — I'd like to make a few comments on the standardization and validation process. Currently, a number of mutagenicity assays have been proposed for use in safety evaluation studies to evaluate chemicals already marketed or being made ready for introduction into man's environment. However, many of these assays have not been validated to the extent where their sensitivity and reproducibility have been established. Further, even the most thoroughly studied procedures, such as the Ames test, have not been analyzed to the extent that positive and negative compounds can be identified routinely by rigorous tests of statistical significance. Also, there are many forms of these assays, depending upon the laboratory involved. Clearly, the time has come to establish procedural standards in the performance of each assay, especially since regulatory agencies are now considering guidelines that may impose requirements for these tests. Criteria must be established which define the validation process itself, and statistical methods adopted which permit all test substances to be evaluated by the same set of rules. Unfortunately, some tests are being made ready for adoption before a critical analysis has been performed on the available literature data. Correlative studies are being quoted in which one or more components lack standards of performance. Because short-term mutagenicity tests hold considerable promise for identifying both the mutagenic and carcinogenic potential of chemicals, it is mandatory that they be developed in a rigorous scientific manner. The good that can come from the application of these assays will be hampered considerably if the standardized validation process is not adopted soon. This process not only validates individual assays, but must involve intertest results and a comparison to relevant human experience when available. When one considers the cost of programs that are being proposed for implementation using these assays, surely the critical analysis suggested here will, in the long run, save millions of dollars and prevent many useless debates over test utility and reliability. A program that includes critical analysis of existing data, establishment of standard procedures, identification of proper statistical models, and requires examination of a minimal number of chemicals of diverse chemical classes for validation purposes should be implemented, I believe, at the international level. This program should be so constructed as to engender participation and support from industrial, government, and academic organizations and have a broad base of information exchange.

4. The Question of Mutagenicity Per Se

Drs. Soares, Sheridan, Hsie, and Krahn made the point that an increase in the mutation rate could have a deleterious effect on the integrity of the human gene pool, with serious consequences for future generations. This is an extremely difficult area because there are no known cases of chemically induced, heritable, human mutations. Whereas, approximately 70,000 Americans die annually from cigarette-induced cancer, it is not known if the children of smoking parents carry an increased number of recessive lethal mutations. What standards should be used in "validation" studies? The response of an assay to known carcinogens has little to do with the use of that assay to assess mutagenic activity. Only a few animal mutagenicity model systems have been developed. Those that are available lack evaluation with large numbers of chemicals. Therefore, one must rely on assays that have a sound theoretical and biochemical basis. Many of these will be limited because they cannot take into account effects seen in the whole animal, such as the inability of a chemical to reach the germ cells.

IV. FACTORS TO BE CONSIDERED WHEN EVALUATING THE RESULTS OF THE TESTS

A frequent problem with the current state of short-term testing arises in the case where data on which a claim is based that a problem exists are thrust upon a manufacturer or user and must be dealt with. In these cases and in evaluating the results of the tests, the following should be considered:

1. Are the data scientifically sound? People are so anxious to report results from these tests that one often hears phrases such as, "Well, it was positive one out of three times." Data must be reproducible. Variable data indicate an inherent problem with the system. The purity of the sample should be known. Appropriate positive and negative controls must be included and should elicit a proper response. There should be a dose-response curve for compounds giving a positive result. A parallel toxicity curve should be run. A great many artifacts can occur as a result of the pattern of toxicity of the compound. These relationships must be understood and examined for any given system. For example, Figure 15 in the chapter by Thilly and Liber illustrates how toxicity can mask mutagenicity. A history of the cell strain should be maintained. Data should be expressed quantitatively. If an acceptable statistical analysis has been developed for the procedure in question, it should be employed. Each of the many tests discussed has its own set of complicating factors that must be controlled and the reader is referred to Section I for detailed discussions regarding specific tests. Dr. Hsie's article is a particularly good example of the many parameters that must be understood and controlled in order to make results from a mammalian cell culture mutagenicity test meaningful. Dr. Brusick's article details the appropriate con-

trols and protocols for bacterial mutagenicity assays.
2. Protocols — Any changes in standard protocols should be noted. For example, incubation of samples for longer than 48 hr in the Ames test results in greatly elevated responses and could be a means of amplifying an effect. Hoever, the background also increases, so that responses are difficult to interpret in the context of a standard Ames test. Dr. Brusick has done additional work along these lines and finds that the treated/control ratio may remain constant with incubation time. However, reproducibility is more difficult to achieve. The point is that such deviations render it essential that the data be analyzed carefully, with an understanding that the altered procedure may yield altered results, the meaning of which may be difficult to ascertain.
3. Activation systems — Levels of enzymatic activity are critical and can vary greatly, depending on the preparation procedure used. A detailed record should be maintained of species, strain, sex, and organ(s) of animals used as sources of S9 fractions and the induction procedures and the manner in which the S9 preparation was stored.

V. POSSIBLE COURSES OF ACTION FOR A POSITIVE RESPONSE

It is clear that a unique set of circumstances will surround each chemical that is subjected to these tests. No set rules could be formulated that would cover the multitude of possible courses of action which might be taken as a result of such tests. Every chemical must be evaluated on a case by case basis.

A positive result in a validated short-term test indicates that there is an increased probability that we are dealing with a potential carcinogen. This assumption should be applied to all decisions pertaining to the chemical in question. Factors to consider and possible courses of action include:

1. Assign someone the authority and responsibility to follow up positive results. The ease of running the tests will result in an increase in the amount of potentially important data that must be handled. A knowledgeable person should be available to evaluate data and aid in formulating a rational responsible course of action for each circumstance.
2. Submit the chemical to other validated short-term tests. Positive results in other short-term tests would tend to confirm that the compound is a potential carcinogen. If the other tests turn out negative, it is possible that the initial finding was a false positive result unique to the particular system and chemical in question.
3. Consider the strength of the positive response as compared to other compounds in the same chemical class. The degree of the response in some tests appears to be loosely correlated with the potency of the carcinogen. However, there are often exceptions. For example, the potent carcinogen dimethylnitrosamine is only weakly mutagenic in the Ames plate test. (It is, however, strongly positive in the suspension test.) In almost every case where a compound was strongly positive in the Ames test, it has turned out to be carcinogenic in animals. Therefore, one should be especially concerned about those compounds that give an inordinately high response in a test.
4. Check to make sure the activity is not due to an impurity in the sample. Sometimes a weak response is indicative of a mutagenic contaminant. If the presence of a mutagenic impurity would have a bearing on the course of action taken, the sample should be analyzed for purity, fractionated, and the individual fractions tested for activity. In some cases contact with the mutagenic component can be eliminated by using a higher grade of material.
5. Compare the degree of structural similarity to known carcinogens and noncarcinogens. If the chemical is related to a known carcinogen, is an alkylating agent, has electrophilic properties, or could decompose or be metabolized to an electrophilic agent by a plausible mechanism, then more concern should be given to the compound.
6. Consider long-term animal exposure tests to determine carcinogenicity and to establish acceptable safe limits of exposure. If such studies are of high quality, then they should carry more weight than the results of the short-term tests.

7. Consider the historical human experience with the chemical and the results of epidemiological studies. In some cases, exposure records have been kept and corresponding medical records can be checked.
8. Evaluate the degree of human exposure to the chemical. If production or use of the material results in the exposure of a large number of people to high levels of the chemical, more extensive testing may be in order.
9. Operations involving the chemical should be examined to see if exposure levels are too high in the light of the new information. Steps should be taken to eliminate any high, unchecked, routine human exposure.
10. In some cases, an acceptable substitute can be found. Sometimes relatively small changes in the molecule may eliminate the mutagenic activity without affecting desired properties of the chemical. One of the strengths of the short-term tests is that they can provide rapid feedback to the synthetic chemist as to the potential carcinogenicity of the various analogues of the chemical in question.
11. Because the data from these assays are so preliminary in nature, normally no public announcement or change in labeling would be in order simply on the basis of the results of a short-term test. However, each case should be examined individually. If results from a reliable battery of tests confirm the positive response, then the appropriate people should be informed.

VI. POSSIBLE COURSES OF ACTION FOR A NEGATIVE RESPONSE

A negative response in a validated short-term test indicates that there is an increased probability that the chemical is not a carcinogen. No amount of data will ever be sufficient to prove that something is not a carcinogen. Nevertheless, the overall picture should be studied, and if examination of the other factors listed above reveals no suggestion of a problem, then the compound can be tentatively regarded as a noncarcinogen and efforts can be directed toward those compounds where a problem does exist. At this time, short-term tests are not replacements for studies involving animal exposure and such work would be necessary to claim noncarcinogenicity of the compound in the context of presently accepted toxicological procedures. The point is that the facilities, time, money, and experts to conduct long-term exposure studies on all compounds are not available, so that some decisions as to the safety and use of chemicals are, in fact, going to be made on the basis of the short-term tests.

REFERENCES

1. McCann J., Choi, E., Yamasaki, E., and Ames, B. N., Detection of carcinogens as mutagens in the *Salmonella* microsome test: assay of 300 chemicals, *Proc. Natl. Acad. Sci. U.S.A.*, 72, 5135, 1975.
2. Clive, D. and Spector, J., Laboratory procedure for assessing specific locus mutations at the TK locus in cultured L5178Y mouse lymphoma cells, *Mutat. Res.*, 31, 17, 1975.
3. Pienta, R., Poiley, J. A., and Lebherz, W. B., Morphological transformation of early passage golden syrian hamster embryo cells derived from cyropreserved primary cultures as a reliable *in vitro* bioassay for identifying diverse carcinogens and recommendations for their use, *Int. J. Cancer*, 19, 642, 1977.
4. Purchase, I., Longstaff, E., Ashby, J., Styles, J. A., Anderson, D., Lefevre, P. A., and Westwood, Evaluation of six short-term tests for detecting organic chemical crcinogens and recommendations for their use, *Nature (London)*, 264, 624, 1976.
5. Legator, M. S., Pullin. T. G., and Connor, T. H., *Handbook of Mutagenicity Test Procedures*, Kilbey, B. J., Legator, M., Nichols, W. and Ramel, C., Eds., Elsevier, Amsterdam, 1977, 149.
6. Legator, M. F., Truang, L., and Connor, T. H., Analysis of body fluids including alkylation of macromolecules for detection of mutagenic agents, in *Chemical Mutagens—Principles and Methods for Their Detection*, Vol. 5, Hollaender, A., Ed., Plenum Press, New York, in press.

7. **Minnich, V., Smith, M. E., Thompson, D., and Kornfield, S.**, Detection of mutagenic activity in human urine using mutant strains of *Salmonella typhimurium, Cancer Res.,* 38, 1253, 1976.
8. **Yamasaki, E. and Ames, B. N.**, The concentration of mutagens from urine by XAD-2 adsorption: cigarette smokers have mutagenic urine, *Proc. Natl. Acad. Sci. U.S.A.,* 74, 3555, 1977.
9. **Weisburger, J. H. and Williams, G. M.**, Metabolism of chemical carcinogens, in *Cancer: A Comprehensive Treatise,* Becker, F. F., Ed., Plenum Press, New York, 1975, 185.
10. **Williams, G. M.** The Significance of Rodent Liver Tumors in Bioassay, presented at the Toxicology Forum, Aspen, Colorado, June 1977, Naylor Dana Institute for Disease Prevention, Valhalla, New York.
11. **Williams, G. M.**, The detection of chemical carcinogens by unscheduled DNA synthesis in rat liver primary cell cultures, *Cancer Res.,* 37, 1845, 1977.

Section III
Discussion of Possible Courses of Action Given Several Hypotethical Test Cases

DISCUSSION OF HYPOTHETICAL TEST CASE A

B. A. Schwetz

TABLE OF CONTENTS

I. Introduction 105

II. Discussion 108

I. INTRODUCTION

This hypothetical case was characterized as a chemical having a high volume of usage, a high degree of human exposure, and no medical evidence of any problems. One good long-term study in rats had been conducted with negative results. Several short-term mutagenesis studies have been conducted with positive results being found. For the purpose of this discussion, the mutagenesis data were defined as positive results in the bacterial mutagenicity test (Ames) and in *Drosophila*. Other mutagenesis tests were not conducted. It was also assumed that acute toxicity data were available, along with the subchronic studies that would normally be conducted prior to the long-term study. The results of teratology and reproduction studies were not available and a pharmacokinetics or metabolism study had not been conducted.

With these available data, the ultimate possibilities regarding the true mutagenicity and carcinogenicity of this chemical in humans are as follows: (1) neither mutagenic nor carcinogenic, (2) mutagenic and not carcinogenic, (3) carcinogenic but not mutagenic, or (4) mutagenic and carcinogenic. Any one of these possibilities could exist but the data suggest that some possibilities are more likely than others. The purpose of any further work on this chemical would be to predict with more confidence or to establish which of these four possibilities represents the real situation.

The range of possibilities of what could be done extends from no additional work on one hand to an extensive investigation on the other, including animal work as well as an epidemiologic study in humans. Considerable discussion evolved around questions such as whether or not these data represented an imminent hazard, whether or not production of such a chemical should be stopped, whether or not the legal departments of companies need to be appraised of such data, whether or not government agencies such as the Food and Drug Administration (FDA), Environmental Protection Agency (EPA), National Institute of Occupational Safety and Health (NIOSH), or Occupational Safety and Health Administration (OSHA) need to be notified and whether or not employees and customers are to be notified. The general consensus was that these data by themselves did not indicate an imminent hazard situation. It would be wrong to do absolutely nothing about these new results, but there are several things which could be done short of considering the situation an emergency. Among the possibilities discussed were the following: additional research to assess the repeatability of the relevancy of the results in hand, discussion of the new data with government agencies, and steps to make the employees aware of the situation and possibly to inform the customers. If the data are to be discussed with government agencies, and if coverage in the news media is probable, the results should be discussed with the company employees as well as customers to minimize anxiety.

Under the Toxic Substances Control Act, does this situation represent a "substantial risk" which must be reported? Again, the consensus was that since additional testing could be done in a very short time to add perspective to the results already in hand, it would not be considered negligent to proceed with additional testing to accumulate more data prior to releasing any results. This, of course, is a generalization that may not be appropriate for all chemicals found to be in this situation. There may be times when this would be considered a substantial risk and it would be prudent to notify the appropriate government agencies. The least defensible choice would be to have the information in hand and communicate the results to no one and take no action to lend perspective to the situation.

If the decision is made to do additional work, the possibilities are quite numerous. Among the choices are

1. Review the medical data and the results of the animal studies.
2. Repeat the mutagenesis and/or the 2-year rat study.
3. Do additional mutagenesis studies.
4. Carry out a 2-year study in a second species.
5. Repeat mutagenesis studies *and* do a 2-year study in a second species.
6. Perform additional mutagenesis studies *and* a 2-year study in a second species.
7. Institute an epidemiology study in exposed people.
8. Do the epidemiology study plus the animal work.

The merits of each of these possibilities were discussed in detail.

A more thorough review of the medical data and the existing results of the animal studies is imperative. The medical data need to be reviewed with the new question in mind: are there any findings in the results of examinations of workers that indicate possible mutagenic effects? If the company is large and particularly if the exposed workers are located at several different geographic sites throughout the country and are, therefore, examined by different physicians at each plant site, it is very easy for isolated occurrences to be overlooked as being irrelevant or spurious. Examination of all of the data with the new objective may reveal some answers that were previously ignored. Regarding the results of the animal studies, it is important to determine if the studies conducted earlier meet the standard of studies currently being conducted in toxicology laboratories. Especially if the study was conducted many years ago, it is not impossible to find investigations that were thought to be "good studies" in the past, but which do not meet today's standards of laboratory practice. Therefore, before additional extensive testing or evaluation is initiated, it is critical to review the existing medical data and results of animal studies accumulated to the present time.

The advisability of repeating the mutagenesis and chronic toxicity studies seems questionable. There may be reasons for repeating the mutagenesis studies — such as knowledge that the test sample was inappropriate or that known positive control or negative control chemicals were not appropriately detected by the test systems or that there were mechanical or personnel difficulties in the laboratory at the time the tests were conducted. If any of these grounds were found to be valid, it would be advisable to repeat the mutagenesis studies. To repeat the 2-year rat study would seem to be of very little help at this point, since it would take at least 3 years to conduct and would not answer the question of mutagenic potential. Thus, there may be reasons to repeat the mutagenesis studies, but to repeat the chronic toxicity study would be a very low priority step at this point.

To conduct additional mutagenesis studies was considered a very logical step. The question was, however, which studies should be done next. A combination of in vitro and in vivo studies was recommended. The in vitro tests might include DNA damage/repair, the mouse lymphoma test, and an evaluation of cell transformation. The most logical in vivo tests included cytogenetics and the dominant lethal test. DNA repair might also be evaluated in vivo. The in vivo studies were considered to be highly critical to define the slope of the dose-response curve and establish a dose level that was not associated with an adverse effect. Such data are needed to facilitate hazard assessment.

To conduct a 2-year toxicity study in a second species was considered an appropriate second-phase step, following the new mutagenesis studies. It is most likely that a carcinogenicity study would eventually be required in a second species, so there was little question about the need for this study. Before starting the long-term study in a second species, however, it would seem advisable to conduct a metabolism/pharmacokinetics study to assess the appropriateness of the species and to help select dose levels.

An alternative to the possibilities already discussed would be to repeat the mutagenesis studies and to do a 2-year carcinogenicity study in a second species. Again, this option assumes that the same mutagenesis tests need to be repeated, rather than starting new mutagenesis tests. The advisability of starting a 2-year study in a second species and the considerations on the timing of this type of work have been discussed already.

The most acceptable proposal would be to start additional mutagenesis studies and a 2-year study in a second species. This combination of work would draw immediate attention to the question of mutagenic potential, would lend perspective to the results obtained in the first studies, and would provide additional information on the dose-response relationship and the evidence for mutagenicity in whole-animal systems. Starting the second 2-year study would also give greater confidence in addressing the question of the carcinogenic potential of the compound.

Consideration should be given to an epidemiologic study in the exposed worker population and other exposed people. The probability that such a study would be helpful depends on several factors. Is the exposed population large enough to permit conclusions to be drawn? Is there a population exposed to this chemical "alone" or were all people exposed to one or more other chemicals as well? Are the levels of exposure documented? Would it be possible to identify an appropriate control population? These and other factors must be taken into consideration before initiating an epidemiologic study. As with all other types of studies, a poor epidemiologic study may be of no help whatsoever, and in fact may do more harm than good. Even if a good epidemiologic study could be conducted, it probably would not be sufficient to resolve the question at hand. It could define with some confidence whether or not a problem existed at the present time within the population of exposed humans, but could not predict the likelihood of problems arising in the same or another population in the future. Therefore, studies in animals should be conducted in order to help predict events that may occur in the future in the exposed people.

A number of factors other than those already mentioned should also be considered in developing an overall program for reevaluation of a chemical such as this hypothetical compound A. A good study of the metabolism and pharmacokinetics of this chemical would seem extremely beneficial to help assess the fate of this compound in several different species, as well as the dose dependency of the fate of the compound. Are there contaminants or impurities in the product which could account for positive mutagenesis results? It may be helpful to isolate these components and test them individually and remove positive reactors from the product. How difficult would it be to further minimize the exposure of people? If the exposure levels are not adequately defined, steps should be taken to get additional information on the levels to which people are exposed. Steps should also be taken to decrease the exposure levels as well as to decrease the number of people exposed to the compound, if possible. Another factor to consider is the availability of another chemical that could substitute for the product being tested. Also, what is the environmental fate of this product? If the compound is one that persists in the environment, the potential problem is much greater than if the product does not persist in the environment. All of these factors, plus others, must be taken into consideration in assessing the ultimate fate of this chemical as a product.

In summary, confronted with the situation of having a chemical that has been in widespread use with no indication of a medical problem, which has been found to be negative in a carcinogenicity study, but now shows positive results in mutagenicity tests, several different approaches could be outlined for assessing the risk associated with the previous and future use of this product. Many factors must be considered in making the final evaluation, but the most crucial input would seem to be the results of further testing for mutagenic and carcinogenic potential of this compound in animals.

II. DISCUSSION*

Compound A — High-volume usage, high human exposure, no medical evidence of any problem, negative in one good long-term rat study, begins showing up as mutagenic in several short-term tests.

Schwetz — I wish I could say that this was just an exercise or a hypothetical case. But rather than asking the question: "if we have a compound like this," it is instead a question of "how many compounds do we have that fall in this category." All I can do is present some issues to consider in this problem. It would not necessarily be helpful to anybody else for me to proceed to describe how Dow Chemical would handle this problem, because it is a situation for each one of the companies to resolve in its own way. So, I would prefer that everybody contribute as we go through this discussion.

What we have to start out with is the information presented: a reliable 2-year study in rats that showed negative results. We have seen some positive mutagenesis data and, just for the sake of a reference point for each of us for the discussion, I've considered that information to be some positive results in microbes plus an effect in *Drosophila*. The other postulates are that this material is characterized by a high volume of usage and high levels of human exposure, or I guess a different way of looking at it would be that many people are exposed to some level of the material. No medical evidence exists to indicate that there really is a problem at this point, so we have some negative data as well as these new mutagenesis test data that are showing up positive.

Participant — How long has this exposure been going on?

Schwetz — Well, probably a typical compound would be one that has achieved a large volume and a large number of people exposed, let's say 10 years.

Participant — Were the mutagenesis tests done in-house?

Schwetz — What you have in hand is a report, either from your own laboratory or an outside lab that says that they have run this test in a standard procedure, let's say two of the Ames strains are positive and two strains of *Escherichia coli* or yeast are negative.

Participant — Does the compound require activation? What other data are available?

Schwetz — Let's say it does. We have the necessary data to precede a 2-year study that will most likely be a 90-day study and some acute work, but that's it.

Longstaff — Let me ask you about the rat study, was it what you might call a well defined group study? You have these negative results, but are you satisfied that it is, in fact, a noncarcinogen in the rat?

Schwetz — That will be at the top of my list of things to determine.

Participant — What kind of use does the material have?

Schwetz — A flame retardant is as good as anything. Typically, you are asking more questions than the information supports. The possibilities I would see for this compound in this situation are

1. It's possible that this compound is neither mutagenic nor carcinogenic.
2. It's possible that it could be mutagenic and not carcinogenic.
3. It's possible that it's not mutagenic and could be carcinogenic.
4. It's possible that it's both.

There are limitations to the interpretation of the data that we have in hand. This compound could end up in any one of these categories.

What can we do to find out which one of those four possibilities represents the real situation for humans? One possibility, of course, would be to do nothing; that would be to assume that it is not a carcinogen, based on that one 2-year study. Another option would be to assume that it is a mutagen or represents a mutagenic potential, based on the work that was done, and to handle it accordingly or to wait for a regulatory agency to decide what the fate of that material should be. So, it would be possible to do nothing about it and just wait. My own feeling is that it would not be a tolerable position. I don't think we could have that information in hand and do absolutely nothing about it. Is there argument with that stand? The

---

* The transcript presented here is an edited version of the original. Some sections have been omitted and wording is not necessarily verbatim. The discussion leader is B. A. Schwetz.

alternatives involve a lot of other possibilities, this is the simplest one.

Henderson — My own reaction would be that one thing you have to do is start informing employees of what you've got. You've got to start your communication to your employees, regardless of whether you do any other testing or not.

Butterworth — This is high human exposure, a lot of people are exposed.

Kilian — What are you going to tell them?

Schwetz — That is a real question. With this information in hand, would you go to your work force internally and tell them of what possibilities exist without telling all the rest of your customers, the exposed people, and the general public? Do you tell the first level and not the rest, do you tell none of them, or do you tell all of them?

Kilian — At that level, I don't think you should tell them, because you yourself don't know what this means. How can you explain to someone else when you don't know yourself?

Participant — I agree.

Participant — So do I.

Schwetz — We are faced with that situation frequently, with having a positive test, and a concern that I've had is that we will be in the situation of calling wolf. If we go out and tell them that these are the possibilities, they never see the rest. They see that there is likely to be a hazard; then when you come back after several months and a couple of tests later and tell them don't worry about it, are they going to trust you?

Clive — There's another hazard about doing nothing and that is that with any popular chemical at all it's going to be tested by someone else. It will show up positive. Your data will come to light that way. The question will be asked ''why did you not do anything with these positive data''?

Schwetz — Yes, that certainly is a reality that happens very often. Fortunately, there are very few instances where the opinion is different than: "Let's do the work and find out if the allegation is true. Let's take care of the problem ourselves rather than have somebody else discover this kind of information and drag us into it second handed."

Butterworth — I think you've said something very important here. Usually we would reason, if it's positive in the short-term tests, then let's do a long-term test to see whether or not it is a carcinogen. The feeling I get from this group is that even though there is one long-term study that indicates that it is a noncarcinogen, the short-term tests are of sufficient value to make you have further concern. Do you agree with that?

Participant — Yes.

Sheridan — Byron (Butterworth), when you say long-term study, I'm making the assumption here that you mean cancer study.

Butterworth — Yes.

Sheridan — We also make the assumption that no mutagenesis study has been done on rats.

Butterworth — Right.

(Sheridan, Soares, and Hsie made the point that mutagenesis per se is also a concern. A compound that is mutagenic in the short-term tests is also a potential human mutagen. This is discussed more fully in the chapter by Butterworth in Section II.)

Sheridan — You said no medical problem, but did you imply that among your workers or the general populace?

Schwetz — I would assume that it means among the workers. The information available to industry is generally derived from their observations on the work force. When we say that there isn't any evidence of a medical problem, it means that based on the usual things that we can look for and see, there isn't any evidence of a problem. That doesn't mean subtle things that would require special additional examinations, looking in detail for some specific problem. So the fact that there's no medical problem evident does not mean that there might not be at some further sophisticated level of detection.

Kilian — I think we can give this a little better perspective. I'm a physician, I've been faced with this problem many times, and you can do a great disservice to your patient by giving him irrelevant information. In the medical profession, if the EKG has an abnormality that we know is not important and doesn't mean heart disease, we don't tell the patient about it because you create tremendous anxiety in these people. So, you kind of have to balance this thing. Of course, you always have the legal people telling you that you have to do everything, but I don't think you'd have the medical people wanting to get involved in this under those circumstances. Now, who's going to tell them if your industrial medical department is not going to get involved?

Brusick — My feeling is that we have at least no overt indication of carcinogenesis and you're trying to use these data from microorganisms and *Drosophila* in a risk assessment for humans; I don't think it's possible to take these data and jump that distance and say that this is either safe or unsafe. I think that there has to be additional information to substantiate activity in a mammalian cell or in a mammal that there might be something. It's just too much to ask to make those kind of decisions from this information other than the fact that there may be a potential problem.

Participant — Under the Toxic Substances Control Act, (TSCA), Section 8e says that we are obliged to recognize, identify, and report a substantial risk. Now, it sounds like the consensus of this group is that we have not identified a substantial risk. Therefore, does that relieve us of the responsibility of reporting that we have found a positive mutagenesis test on microbes and *Drosophila*?

Ray — I was just thinking that at this point you know you have a substance that is active in microbes and an insect, you've got the option of doing some other tests that should not take much time. You have the negative rat study. It seems to me you've got enough information to hold until such time as you get more information. Its relatively a short time frame.

Legator — I agree with Verne (Ray) that we're really talking about something now that can be done relatively fast. One of the things that is bothering me are the proposed new EPA guidelines for mutagenicity testing of pesticides. Would the positive *Drosophila* and the microbial test be enough by their present document to call this a mutagen?

Flamm — Well, as far as the proposed guidelines for Section 3 under the Federal Insecticide, Fungicide, and Rodenticide Act (FIFRA), it is not yet clear. Let's assume that this is not a pesticide, and let's go back to Section 8 of TSCA that was just mentioned as to what industry's responsibility is. The wording is unclear as to what constitutes a substantial risk. The wording is unclear as to what constitutes an unreasonable risk to human health and the environment. It won't become clear, in all probability, until they begin to promulgate regulations under Section 4 which have to do with standards for protocols, wherein it is mentioned and provisions are made for mutagenicity testing. So it's Section 4, promulgation of TSCA, which will help to define what is meant by substantial risk under Section 8. You brought up the FIFRA guideline and criteria activity and it's good that you did because it's my understanding that currently the agency plans to use those proposed regulations, guidelines, and criteria for implementing Section 4 of TSCA. So, that could apply and would help to define what is meant by substantial risk. At least that's my understanding of things as they are now. That could easily change.

Participant — There is quite a possibility that a private organization may make the same findings. In that case, if you haven't considered discussing this matter with the Agency in some form, you are in the position possibly of having obtained information and not having revealed it. Particularly if there turns out to be some public concern over the findings by the private organization. I think substantial risk indications here need very careful exploration in order to decide what needs to be done next.

Schwetz — I'm sure the worst situation would be to have this information in hand and not talk to anybody on the outside in addition to not doing anything inside.

Participant — Is it, therefore, the consensus of the group that in this situation you don't tell EPA or your employees?

Kilian — If I had the decision, I would inform NIOSH of the tests and also let them know that we were doing more in this area.

Schwetz — I think there is a difference between telling them there is an imminent hazard and discussing the data with them.

Kilian — We inform NIOSH a lot of times of what's in the mill.

Schwetz — But we don't necessarily tell them that this is the result we got yesterday and we think it's an imminent hazard. There are ways of letting them know about these things without creating a panic.

Golberg — Aren't there two parts to this? One is to inform NIOSH, but at the same time to tell them what you propose to do next, rather than wait for them to tell you what you ought to be doing?

Schwetz — Well, there are obviously two ways of doing that and even within a company there are proponents of each side. There are some groups who want some agency to tell them what to do, so that when we do it, we're sure that they will accept it. There are many of the rest of us who feel we do not need to go to them for their blessing; let's do what we think is right, then go back and discuss the results with them and come to some understanding.

Legator — Let's say we communicate with NIOSH. Having once done that, then the chances of it getting out to the lay public are fairly good, especially if it is a toxic chemical.

Ray — You might even say high.

Legator — Now, that brings me back to the original question about informing the employee. As a strategy, we agree that it is not a proper thing to do, because we're really not sure we have anything. Once we've communicated to NIOSH, our employee may find out by indirect means that he is exposed to "a hazard." Is that a real problem?

Schwetz — Anytime we go to the outside with this kind of information, we simultaneously tell our people internally because the worst situation again is for them to read it in the newspaper.

Butterworth — What about your customers?

Schwetz — If it's a kind of product where we have a number of customers who can handle that kind of information, that would be relayed to them also, at least with the fact that this information exists and has been transmitted to NIOSH.

Legator — I really question whether you would go at this point. My feeling says you go for another month or two to do something in animal studies or other tests and then go when you have a more complete picture. We are talking about such a small difference in time here that I don't think you would really be guilty of withholding information at this stage.

Schwetz — If the decision is made to do something, these are among the possibilities that I would see: one is to review the medical data, and also review those long-term animal studies to see that the first impression was right; the next possibility would be to repeat the mutagenesis and the 2-year study; do additional mutagenesis studies; do a 2-year study in a second species; or do epidemiology in exposed people. What I would like to do is go through each of these options individually and discuss specific points. I saw as one of the most critical things, the need to go back and review the medical data as well as the 2-year animal studies. Fairly often, that first impression on the medical data is accurate, but if you have a large company where there are a number of people at several plant locations being seen by several different physicians, while the overall opinion and the first glance might be that there is not a medical problem, it's advisable to go back and look at the records in more detail to be sure that that's a firm conclusion that you can live with. This same thing applies with the animal studies. Any of you who have had the occasion to go back and audit a 2-year animal study that was done 10 or 15 years ago know you may find that either it was not quite as good as we thought it was or that it was good for the time but does not necessarily allow you to draw the same firm conclusions that we can from a 2-year study that we're designing today. So I think it's critical to go back and review those before you make major conclusions on whether or not that test was valid.

Golberg — One of the big problems very often is that the dose that was used in those days didn't match up to a maximum tolerated dose under present standards.

Schwetz — We have had studies in both directions. We have had some where the high dose levels exceeded what we would consider the maximum tolerated dose today and some that were less.

Participant — Another point in evaluating data is that you are frequently looking at things within a normal range and if you have an indication that there might be a problem, then those things that are at the lower or upper end of a normal range tend to get a little more important.

Soares — Would these medical data include some kind of review of possible genetic lesions in people exposed to the compound in question, such as fertility data or spontaneous abortions in women?

Schwetz — That would vary from company to company. The information would be valuable if it were available. Often, you may not have that history going back very many years. That was not the first thing that was being reported years ago.

This gets us into the question of what do we repeat, if anything. There may be reasons for repeating the mutagenesis studies. Maybe, in looking at it, you found that the positive control didn't really look like it should, or the negative control wasn't quite as clean as it should have been. Maybe the sample that was tested had a much higher level of stabilizer in it than is typical of production lots of the compound. Maybe you should go back and look at the stabilizer. One concern of mine is that we tend to consider repeating things more often when they're positive than when they're negative. That is a form of bias we should not accept.

Schwetz — What about repeating the 2-year rat study? There are some that we have looked at where it would be advisable to go back and repeat them. Let's accept the condition that this was a good 2-year rat study, and under that condition I don't think it would be an advisable use of resources.

Participant — One of your early considerations should deal with the chemistry of the compound as to whether you would expect it, from its relationships to other compounds, to be active or not.

Schwetz — That's obviously an important point.

Participant — At this point it is very likely that we would have a reproduction study or a teratology study or both to fill in.

Participant — I doubt sincerely whether or not you would have any idea of a heritable component to this kind of activity.

Schwetz — One possibility, rather than repeating those same mutagenesis tests, would be to proceed with additional mutagenesis studies. I hadn't intended to define which ones at this point, except to assume that this would be the next level of testing, more detailed, more sophisticated tests than we have already done, but not necessarily the specific locus test or some of those that would represent the ultimate tests.

Butterworth — I think this is a very important question. We agree more short-term tests are needed. Which tests would the group choose at this point to supplement the bacterial and the *Drosophila* tests?

Legator — Obviously, you would want to do in vivo cytogenetics. I might want to do mutagenicity in blood and urine, DNA repair studies, and the dominant lethal test. I think the cardinal mistake that was made right at this point, that got us into this position, was the failure to conduct those tests concurrently with the *Drosophila* and microbes. We should have been doing a battery of tests, including animal tests, so that we would have a definitive look at this compound at this stage; I think that would have eliminated a lot of problems that you are now discussing.

Participant — In vivo cytogenetics is not very sensitive. If you look at in vivo cytogenetics, it does very poorly on compounds that require activation, for example, acetylaminofluorene and dimethylnitrosamine.

Soares — In addition, natural insults such as viral infection can result in chromosomal abnormalities.

Legator — I think now we have to build up a data base in animals and work back and forth with people if at all possible, but I think right now we have to plug in some animal studies. Another important thing is to get an idea about metabolism and pharmacokinetics.

Schwetz — If you go the in vitro route, it would not be as easy to get a dose-response curve that we can extrapolate to the exposed people, as if you go the whole animal route. Does that make a difference at this point as to which studies we would recommend or would we simply go to both?

Butterworth — I think it would be very important to look at a cell transformation assay, at mammalian cell mutagenicity, or DNA damage.

Brusick — If one goes directly to animal studies, it is hard to establish dose response in dominant lethal or in vivo cytogenetics. A compound may not produce significant chromosome aberrations, it may not turn out to have any response, in which case you would still have to go back and work with mammalian cells to see whether or not you can find gene mutations or chromosomal aberrations.

Legator — We can do many in vitro tests, I think they are all great, but the critical point we're coming to is we have got to start to develop some animal data and I don't see how we can escape it at this point in time. I would again suggest captan as an illustration of a widely used compound that gives us all sorts of positive information in vitro, but yet it is totally lacking, as far as I am concerned, in any sort of information in vivo, and I think here it's because of a highly active metabolite that is produced probably without problems. Strategies without essential animal data would have lost valuable compounds, such as captan, and I am sure many others.

Swenberg — I hear a suggestion that we do both, that it is critical to get into animals at this point and not just do in vitro work but also to do in vivo as well.

Thilly — It may be possible to look for parent compound or metabolites in exposed workers.

Schwetz — If we have people out in the plant exposed to this material at a level that perhaps would permit you to find some metabolites in the urine, or you could take a blood sample and measure metabolites or parent material in those people; that would be helpful information in looking at the results of pharmacokinetic studies to see if the same metabolites are formed in mice that are formed in humans, but not formed in rats; that might be an indication that the first rat study was not appropriate. So, if that information could be gotten, it would surely be helpful.

Ray — You know what the parent compound is and if you have done some metabolism studies on your own, you would have had some appreciation of the possibilities of metabolites that you might find in man and you might even have a chemical assay to identify the compound(s). Now you can look in man using "cold" material, but to go the route of radioactive compounds in man usually is not practical.

Participant — That can be a major undertaking, trying to develop the analytical techniques.

Ray — That's true.

Participant — If you are talking about low-level exposures, you would have a hard time demonstrating metabolites in blood or urine.

Schwetz — A point that is mentioned in the proposed pesticide testing guidelines is that if it can be demonstrated that the chemical or its metabolites are not present in germinal tissues, that will also help to lend perspective to those positive results. How do we analyze and what do we analyze for in the germinal tissues to reverse that? You are limited to a level of detection, there isn't anything we can absolutely say is not in the testes. So is that really helpful for us or not?

Flamm — I think the hycanthone experience could help here because they did have the data on its distribution and it was a surprisingly small amount that reached germinal cells. It was, according to my recollection, 100th or less of the peak blood levels. Using this information, together with the dominant lethal test and test information from 12 different test systems, it was possible for Abrahamson and others, including myself, to calculate that the level of risk from one administration of the compound in humans would probably be of the order of 1 to 5 rd of radiation. So, of course, this being a therapeutic compound used for the treatment of a very serious disease, you have a different kind of risk-benefit consideration. But I think, nevertheless, it might be germane to look at that example and to see how we did ameliorate the fears of genetic hazard and risk.

Schwetz — It is a good example. It would be very helpful if we could get that information.

Flamm — That's available in a symposium document.*

Soares — If you are dealing with a mutagen, there is a chance of progeny being affected. So you have to move quickly, you cannot wait 3 years to do a complete pharmacokinetics study.

Participant — Suppose we do in vivo cytogenetics, DNA repair, in vitro mammalian mutagenesis, and a dominant lethal and they all come out clean. Where do we stand then? We still have the original results and now we have all these; what do we do now?

Legator — I think that if everything is negative, that you just read off, plus the negative carcinogenicity study, we would be fairly safe. However, the one piece of information that I would like to fill in here someplace along the line would be to find out why we had the discrepancy.

Participant — Suppose it shows up as positive in some of these other tests such as in vivo cytogenetics?

Legator — I think that we would then call that compound a hazard. I cannot see where we would not call that an active compound.

Schwetz — Also, in the process of making that decision, it would seem prudent to look at some of the people who are exposed.

Schwetz — Doing a 2-year study in another species, would be a possibility, but would it

---

* *J. Toxicol. Environ. Health* 1, 173, 1975.

help this situation? Because the question is one of mutagenesis and we may confirm that it is noncarcinogenic in the second species and 3 years later we're still debating the mutagenesis question, so it is not a helpful thing for us to do all by itself at this point. If we do additional mutagenesis studies, do we want to do a 2-year study in addition to that?

Participant — In a second species?

Schwetz — Yes. To start a 2-year feeding or inhalation study, in mice or hamsters, on the basis of the rat data is risky because you can get 6 months out and find that maybe the males respond differently than the females in mice and you didn't anticipate that or they are less sensitive or they are more sensitive.

Legator — If you are going to take a compound out of the marketplace based on positive findings in animals in terms of mutagenicity, then you might be ill advised to go ahead with that 2-year study because you're going to make a decision at a much earlier date, on the basis of the mutagenicity data.

Swenberg — We are all learning about short-term tests and I think we need the correlation of that second study. This is a high-volume compound, and it's obviously a big market; the company cannot afford not to do the second carcinogenesis test.

Legator — If you can afford to do the study for the sake of the correlation, that's great. I think here you have to ask yourself, are we talking about an area of toxicology in its own right, mutagenicity, or are we still using this only as an indicator for potential carcinogenicity? If you make a decision that we are talking about an area of toxicology in its own right, then you must realize that the long-term test will not have a bearing on the mutagenicity question.

Schwetz — If the red flag is going to be raised on this compound in the immediate future, it's going to be on the basis of the mutagenesis data. In order to give this compound protection long-term, we will need the second species 2-year study anyhow. We are going to get the mutagenesis data much more quickly. It's going to be at least 4 years, from the time we first write out the protocol for, say, a 90-day study in a second species, get that one done, and get a protocol written for a 2-year study, which then takes 3 years. We are talking pretty close to 4 years down the line, and I see that as insurance data. Whether or not we bring the red flag down or the yellow one is going to be based on the next 6 months of mutagenesis data.

Participant — I would say that you would start your 2-year approach immediately and get it going and then actually if the axe comes down, you would terminate without having really spent your $150,000. You're at least time ahead if the mutagenesis work comes out still a little fuzzy; you have at least got whatever months you have well under way.

Schwetz — Generally, it's a question of resources that you can put into that question right now. If I sense some of the feelings here, the immediate issue and, therefore, probably the need for the immediate money and space would be to resolve that question of mutagenesis.

Participant — At this point I would check to see if there were unnecessary exposures that can be curtailed or minimized. The whole process of manufacturing should be checked at this stage and if there is suspicion of unexpected or previously unknown concern, then measures should be taken, that are feasible without redoing the whole process, to minimize exposure at this early stage.

Participant — I would like to go back to my other question again: if we did these five tests and they were all negative, would we stop there or do most people feel that if they all come out negative then there is no problem, or would we still have to do a 2-year study and spend another $200,000? I look at this compound as something that the studies have been done on while it's on the market and now we are coming back to the reregistration and new requirements doing mutagenicity work in retrospect.

Participant — I think when the mutagenicity test protocol was first proposed, you should have asked yourself the question, what do I do if I get a positive result and what do I do if I get a negative result?

Participant — I think you have to answer why were they positive in one system and not in the other?

Participant — Do you do that before you do your second long-term test?

Participant — I would like to just make an attempt to answer that question, from my point of view as a student of toxicology. I think safety evaluation has to be a continual concern and something you have to do on a day-by-day basis. I think if you get all negative results, what you have to do is say, maybe for this situation it's not quite as high a priority item as something that has one or two positive findings. You also have to justify this from a cost point of view. I would say that you are going to have to address this sometime in a mammalian species and try to get as close to man as you can. I feel strongly that a positive finding is only a red flag, and I feel the same way with a negative result — it's only tentative.

Swenberg — In response to this point, if these other tests are negative, the time involved in getting the results of these negative tests provides an opportunity to do the subacute study in a second species. Current guidelines clearly recommend carcinogenicity evaluation in two species, not in one. Therefore, I do think you clearly have an obligation to go in and be looking at the compound in a second species. You probably won't be ready to do that until you have your results from these additional mutagenesis tests. If, indeed, you want to kill the whole product, maybe you wouldn't do it. If you have any idea that you might keep that product around, you are clearly obligated to go in a second species.

Schwetz — I want to look now at the question of epidemiology. That, obviously, is one thing that could be done. I am not an epidemiologist, but I think I've been close enough to some of these studies that have been run to know that it is not nearly as easy as I may have thought a few years ago. It is very difficult to find a population of exposed people where you can document the exposure and know what they were exposed to. It is just about as difficult to find a control population for those exposed people. It is difficult to go in and get the right questions asked and to get the right answers to those questions. At any rate, if you did a good epidemiologic study, would that be sufficient to negate everything else that's going on? No, it would not. So, this by itself, I don't see as an answer. Again it might be helpful information. On a long-term basis, maybe an epidemiological study would be advisable but it's not the first thing that we would do.

Participant — At this point in time it would be important to identify a cohort of people.

Schwetz — That is true. It might be the time to step up your observation of those people. I just want to reiterate that we have looked at what animal work we can do and what human work we could do to help resolve the situation. But there are other things that would be good to fit in between and, obviously, metabolism and pharmacokinetics would be helpful to find out if the 2-year study in rats was the right choice. If we are going to do a second species, this might help in choosing which one. Considering the dose levels used in the rat 2-year study in conjunction with this metabolic work is valuable. Perhaps there's a metabolic threshold just above the dose levels used in the rat study that would explain different kinds of results in other test systems that are exposed at higher dose levels. So metabolism and kinetic data would be helpful in making the decision on what to do next.

Is this compound one that is registered or is it something that will be registered? Is it a contaminant in another product? For some products, the 2-year studies in two species are going to be required. We've already touched on the questions of contaminants or impuries. George (Levinskas) brought up this question of the ease of minimizing exposure to people; we'd better take a look at that right away. If the operation can be "buttoned up" so that we no longer have exposure, we've already eliminated one of the two parts of the hazard. The question of toxicity may still exist but we've eliminated or significantly reduced exposure so the hazard is reduced. That provides some time then to help assess the toxicity. At this point, with this much investment in the product, it's not likely that the company has a number of other alternative compounds that can be substituted, but that is a possibility also.

Schwetz — Another point would be the environmental fate. If this compound is one that persists in the environment for long periods of time, leaving it on the market will result in a worse problem 5 years from now than it is right now; that's different than if it is biodegradable and is not an escalating problem on a long-term basis. Here we have a further factor to consider in looking at the overall picture.

Henderson — In regard to environmental fate, I'd like to go back to the comment that Dr. Legator made that mutagenicity is an area of toxicology in its own right. Should we be concerned with putting into the environment materials that will increase the rate of mutations of microbes and insects, because we have tests indicating that the compound will do this. We've been thinking primarily in terms of taking those data and applying them to man; we've forgotten that they tell us something about what happens to microbes and insects in the environment.

Soares — When you speak of plants, insects, and organisms other than humans, you have to keep in mind that natural selection is an extremely powerful force. Humans are, in many respects, immune to the effects of natural selection. For instance, a phenomenon such as Down's syndrome would not survive under natural conditions, but in human society it is going to cost us a quarter million dollars. But if something like that were to occur in *Drosophila,* natural selection would work against it.

DISCUSSION OF HYPOTHETICAL TEST CASE B

V. A. Ray

TABLE OF CONTENTS

I. Introduction 117

II. Discussion 118

I. INTRODUCTION

The situation described for Compound B is one that is going to occur with increasing frequency. Compound B is a postulated development candidate; in this case it is assumed it is a potential oral drug to be given chronically. The structure is unique, so there is not an existing frame of reference to aid in predicting mutagenic/carcinogenic potential. The compound is potentially valuable, but gives a positive result in the Ames test that has been set up as an early screen. The question becomes: "do we drop the compound at this stage simply on the basis of a positive Ames test"? The consensus was that the compound deserves further study and that it would be unwise at this time to make any decision based on the result of a single test. However, the amount of testing to negate the positive Ames test would not be trivial.

For the purposes of this discussion it was assumed that compound B was active only in strain TA100 and negative in the other Ames strains. It was also assumed that the compound did not require metabolic activation. The finding that a compound would be positive in one base pair substitution strain (TA100) but not in another (TA1535) is disturbing and inconsistent. However, participants pointed out that just such cases do occur.

In order to better define the genetic activity, it was suggested that other genetic assays be used. Those that might be available within a drug firm and which were proposed were forward mutagenesis in bacteria such as the *Escherichia coli* system, mutagenesis in mammalian cells such as the mouse lymphoma model, assay of the urine of the animals administered the compound in the Ames strains, mytotic recombination in yeast, and in vivo and in vitro cytogenetics. A positive result in any of these tests would lend support to the suspected genetic activity. It was felt by most that negative results in all of the above-mentioned tests would be adequate to reduce the impact, if not negate the positive result in strain TA100. Once again, it was the experience of some investigators that just such results had been observed.

Another approach that might be followed would be to examine analogues of compound B for their ability to mutate TA100. If an analogue were negative in the Ames test and retained all the desired pharmalogical and pharmacokinetic profiles to merit development, it would be prudent to develop this compound rather than compound B. Suppose, however, that no such compound could be found. We are at the stage of wanting to submit an Investigation of New Drug Application and proceed with clinical trials: Phase 1 — human tolerance and Phase 2 — evaluation of efficacy. The question of genetic activity has been resolved by further mutagenicity testing. However, the correlation between Ames test results and carcinogenicity requires that the question of carcinogenic potential be addressed. What can be done short of expensive lifetime animal exposure studies to shed some light on this problem? It was felt that mammalian DNA damage and malignant cell transformation assays could be employed at this point to provide sufficient assurance of compound safety to proceed to Phase 1 clinical trials.

It was the feeling of many participants that preliminary results from cell transformation assays indicated that they would be particularly accurate predictors of carcinogenic potential. Cell transformation assays will be used with increasing frequency to clarify such data. However, it was also pointed out that these tests require substantially more development and were among the most difficult to perform. A positive result in such a cell transformation assay at this point would probably be sufficient to stop development of compound B.

II. DISCUSSION*

Compound B — Development candidate; potentially valuable new chemical; unique structure; positive in the Ames test.

Ray — The nature of the data provided affects the manner in which we would proceed. I'm going to try to set up what I consider to be the most direct kind of analysis. Let's see what the group does with it. What we're concerned with are the consequences of a positive finding in an Ames test. Let's say for the purposes of this example that the compound is reproducibly active in TA100 and negative in the other Ames strains and that the compound did not require metabolic activation; in other words, it was active directly. Let us also postulate that this compound is a potential antiarthritis drug intended to be given orally. This means it would have a chronic exposure pattern to it. It also means it could be used in adults of child-bearing potential; in other words, have a potentially broad usage. The time of finding out that this compound is an Ames positive is today. What would be our next step?

Participant — Where are you in your toxicology?

Ray — We do not have a 30- to 90-day study. Let's suppose we chose to go on now to find out something more about the genetic activity and chose to assay in the following systems: a forward mutation assay such as an *Escherichia coli* system, using the galactose locus; the point mutation in the mouse lymphoma model, using a TK locus; assay of the urine of the mice administered the compound by the Ames strains, obviously including TA100; a yeast D3 and D4 assay to look for mitotic recombination and mitotic gene conversion; and we did in vivo and in vitro cytogenetics. Now, let's say that all these tests were negative, in other words, we still have the single positive result in TA100. The compound is producing mutation in a base pair substitution mutant and has no apparent effect at the chromosomal level.

Participant — I would check this thing for a false positive.

Ray — It is possible. Let's say for these purposes that you did that and you had no evidence that the compound was affecting histidine metabolism in the strain. Now, what would you as a toxicologist advise next? What are the possible courses of action that you would take? One course of action, clearly, would be to try a series of close analogues. You are in a developmental phase, and although the pharmacologists tell you that its the hottest thing they have seen in years and is the best drug for arthritis since DMSO, nevertheless, you have to map out a plan of action here that would try to identify a close analogue with the desirable pharmacological activity which did not have this observed genetic activity. This strategy presupposes, of course, that such analogues have the necessary pharmacological profile to indicate that they are potentially valuable in the disease condition mentioned. Now, if one of these does not have the capacity to mutate TA100, then the prudent

---

* The transcript presented here is an edited version of the original. Some sections have been omitted and wording is not necessarily verbatim. Discussion leader is Verne Ray.

course of action might be to develop this analogue, even if its potency were less than that of the original compound. If, on the contrary, the original compound was the only member of this series which had the proper pharmacological and now pharmacokinetic profiles to merit development, how would one proceed? Would it be advisable to discontinue any further work on this series? Would it be reasonable to presume that the only way one could effectively deal with an Ames- positive result such as this would be to go directly to a long-term rodent study? After all, it is a chronic-use drug. In other words, wait 2 years or so before proceeding to clinical trials? This doesn't seem to be a logical way to proceed to me. Any differing opinions?

Golberg — Yes, I think one point you overlooked in the previous alternative is that there is always a stigma attached to an analogous compound, even though it might be negative in the TA100. The fact that there was an analogue that was positive makes it difficult to allay suspicion. I can think of somewhat different instances, for example, where a member of the new class turned out to be carcinogenic in rats. For all analogues of the compound there was always a worry, even though the long-term studies came out negative.

Ray — In this particular case, there was no close analogue that had the pharmacological efficacy profile so we're going, at this point in a hypothetical case, with the original compound.

Participant — I think here again, as in the previous hypothetical case, you would like to know the mechanism of action and why the compound is positive in TA100.

Legator — I think you've taken a little bit of the sting out of a positive TA100 because you have now buttressed it with so much more information, with recombinants, forward mutations, and other test systems including cytogenetics. The other negative test results have now been buttressed to the point that we would strongly suspect that this may be active simply in the TA100 tester strain.

Thilly — What I'm having difficulty with is the fact that it gets to the joints, suggesting that it's reaching the urine. If it were appearing in the urine, obviously it would be positive in the TA100 test, because it is positive in the TA100 test as the original compound.

Ray — It could be metabolized in vivo so that you no longer have this mutagenic activity.

Clive — Verne (Ray), I have a feeling that you've worked us into a square circle here. Is this just an academic exercise?

Ray — Be assured, it's not.

Clive — Then I think you have to go back and find out why it was positive in the TA100 strain; do a little basic research on the problem.

Ray — How would you suggest we do that? Would it be reasonable to presume this course of action to be feasible? Mind you, the compound has to go to human toleration studies and a determination of efficacy in human beings before we know whether it is really active in man. So we're really hung up here to get to clinical trials: Phase 1, which is human toleration, and Phase 2, in which we have some idea of efficacy. On that basis, let me suggest another strategy. The components are as follows: we would assume that the effect of TA100 is unique to that strain on the basis of the negative results in all the other assays employed. But we would still have to deal with a correlation of a positive Ames test with potential carcinogenicity. Do you feel that's true?

Brusick — That's the point I was wondering about. Are you concerned at all with the view of the Food and Drug Administration (FDA) towards submission of an Investigation of New Drug Application (IND)? How they would react to this, because I know that there are situations at FDA where they do have positive *Salmonella* data on a compound or on compounds that are already being used on a chronic basis and, as far as I know, no action has been based upon a unique microbial response.

Swenberg — This is a compound that has shown effectiveness. I would suggest that you might be able to get 2-day clinical trials.

Ray — Well, that's the objective you see: to proceed to human toleration studies, and in doing so take blood and urine and so forth from the few patients to whom the compound is administered, so you'll get some idea of metabolism as well and all the other things you try to do in a Phase 1 study. But you have to get there. You have to have an IND application approved. Let's assume that we would still have to deal with the correlation with carcinogenicity. From the mutagenicity standpoint you've done the reasonable battery of tests, but we really worry about the correlation between Ames tests and carcinogenicity. Are there other assays that we could run, which would give us perspective on that? I propose the following: could we run a DNA damage assay and an in vitro transformation assay to provide sufficient assurance of compound safety in order to proceed to Phase 1 clinical trials? Here, let's suppose we had negative results in these additional assays. Would we, in the view of this group, now have enough assurance of safety to go to man in clinical trials intended firstly to assess safety in man and secondly to determine efficacy?

Flamm — What kind of DNA damage studies?

Ray — Well, it wouldn't be the microbial ones. I would say we would do studies in mammalian cells for DNA damage and unscheduled DNA synthesis.

Flamm — Well, following along the suggestion that you might try to determine what happened in the TA100 system where you were getting the unique response, you might want to look at that DNA as well to see if you can measure anything, and with equivalent sensitivity look at systems that would seem to be more relevant and cogent to the safety issue.

Ray — Do you feel that running an in vitro transformation assay here provides sufficient assurance to get into man, considering the battery of test results that you now have? I feel that it is a role really for an in vitro transformation assay. I think cell transformation assays will be focused on more and more to clarify data, such as that I've presented. If you have only an Ames positive and you've done a reasonable job on the tests for mutagenicity, is it a plausible position to run an in vitro transformation assay, and does a negative outcome provide sufficient assurance for safety to permit early clinical trials?

Golberg — I feel we have to bite the bullet and if we do this in vitro transformation test we must have confidence in the results.

Flamm — I would want to see it done with appropriate positive controls, so that if the results with the test compound were negative there would be assurance that in that laboratory you do get an appropriate positive response with positive controls that are reasonable and appropriate in regards to the test substance.

Participant — I feel it is important to choose the positive and negative controls carefully so that they, as much as possible, reflect the structure of the compound. The test should be valid for the class of compounds one is interested in.

Pienta — I think this question of validation of short-term tests is very difficult to answer. I was told to come up with a standard test that people could use. First we had to decide which chemicals we were going to test our system against in order to validate it. When you go back to the literature, you find that many carcinogen bioassays were not done very adequately in the animal system. Our primary standards really aren't 100% valid. I'd like to know what animal system there is that will detect every carcinogen hazardous to man. You can't do that in the first place. There are many questions that the clearinghouse is faced with in establishing the proper animal bioassay systems. Now, we're trying to match our data against these systems and it's an awfully difficult situation. I put down numbers as everybody else, such as a 98% correlation or a 73% correlation. Many of those claims are meaningless because you can select your data so that you can obtain 100 or 0%. I think what we really have to do is decide what chemicals can one test and prepare a compendium of chemicals tested by all these systems and then examine how they really match up. We could be doing this for the next 50 years. I don't really know what you do. We call them as we see them. If they transform cells, that's what we say, they transform cells. I think that it gives us a flag to say, well if they do transform cells, then we should consider further testing. I don't think there's any one test we can use. We've looked at about 100 compounds and we've matched the ones we've tested against the ones Ames has tested and there are some carcinogens that we miss, there are some that Ames misses, and there are some that we both miss. I think it's awfully dangerous to put all your eggs into one basket with a one-tier test. I think we're forced to use a battery of tests. I think the decision that needs to be made is which tests should be included in the battery.

Williams — Verne (Ray), in the situation that you've constructed here, getting a positive result in the TA100 without metabolic activation, you presumably have an agent that if it's active at all, is direct acting. So, when you go to a transformation system, the biggest shortcoming has already been eliminated, that is, the need for metabolic activation. This is where most transformation systems fall down, but as regards to direct-acting compounds, the track record is pretty good. If you had performed a couple of the standard transformation assays and obtained no positive results, I would say that that is a strong indication that your original result was an artifact of some sort. Now there's one other possibility that would have to be considered and that is that the compound actually underwent a metabolic transformation by the bacteria and, of course, there are compounds like that, cycasin being an example. So the structure has to be examined for the possibility that the bacteria were, in fact, activating the compound and that's why it was negative in the mammalian systems. But I would think that if that were properly excluded, that you've got a situation indicating a pretty safe compound.

Legator — How about the other situation? Suppose we were to get a positive result with this cell transformation test and everything else were still negative; would we be ready to give up on that chemical at this point?

Ray — Yes, I would say that would be a pretty big hurdle to crawl over.

Swenberg — Verne (Ray), I think everybody seems to agree, if all these other tests were negative, probably the result in TA100 is not all that meaningful. The problem is the political one rather than the scientific one. I would guess that you'll be limited. If you're lucky, you'll get an IND for limited clinical trials, but you will probably have to do your carcinogenicity studies early before you get into long-term trials. I think that's reality. I could cite numerous examples where far less hazard was indicated than the positive Ames test, in which trials were limited to very short duration.

Ray — That's one of the reasons I set it up as a chronic-use drug, because it's pretty clear that you probably have to go to long-term studies.

Flamm — Forgetting about the political aspects, the one thing that could be alleged that you might have to address is whether you're really comparing equal things from the standpoint of sensitivity. Are you moving your baseline up and down as you go from one test to another, which may qualitatively look at the same things but quantitatively be very different? This is what Ames so often says when things are positive only in his tests. He says: well, it's because my test is so much more sensitive than other tests. But there are ways of doing the necessary dosimetry so that you can construct the situation of comparability with respect to the quantitative issues.

Brusick — Well, I think I would tend to go along with the adequacy of the tests that you've run. They would certainly reduce immensely the impact of that positive result in the Ames test. Also, with respect to elucidation of the mechanism, you do have some insight already in view of the fact that TA1535 (which is TA100 without the drug-resistant episome) yields a negative result. You can ascribe the response to the difference between TA1535 and TA100.

DISCUSSION OF HYPOTHETICAL TEST CASE C

M. Hite

TABLE OF CONTENTS

I. Introduction 123

II. Discussion 124

I. INTRODUCTION

I have been charged with a discussion of "Compound C." The information provided on this compound is as follows: intermediate in a plant production step; modest exposure only to workers in the plant; related to a known carcinogen; negative Ames test; negative Drosophila; negative dominant lethal; positive mammalian cell transformation assay. I might add this is not a Merck compound. I will assume that each of the above tests was done in a scientifically sound manner and that the data are valid.

As presented, this particular compound was negative in three tests for mutagenicity, namely the Ames test, *Drosophila* assays, and the dominant lethal test. These tests are known to measure only mutagenic events and are only presumptive for carcinogenicity. In addition, the data on Compound C show that it was positive in a mammalian cell transformation assay. This was the only test specifically designed to assess the potential carcinogenicity of a compound.

In my opinion, the three mutagen tests should be done using the chemically related known carcinogen. It also would be worthwhile doing a DNA repair test (such as the alkaline elution assay), the Rosenkranz test, or an in vitro test for unscheduled DNA synthesis. These tests should be done with Compound C as well as with the chemically related carcinogen. Information gleaned from these DNA tests might help in the evaluation as to whether or not Compound C has carcinogenic potential.

Of course, the purity of Compound C should be established. In addition, metabolism studies of the known carcinogen and Compound C would shed light on the biological disposition of each. The above discussion is concerned with Compound C and a known carcinogen to which it is chemically related.

Given the positive result of the cell transformation test, it would be prudent to require the workers to wear proper protective clothing. In addition, the environment should be periodically monitored to measure the concentration of Compound C in the air. Also a medical surveillance program should be instituted after discussion with the plant physician and proper engineering controls should be put into effect. The people responsible for safety should write precautionary statements. The chemists should seek an alternate route of synthesis to avoid the use of compounds with carcinogenic potential, and the final product should be analyzed for the presence of Compound C.

In my opinion, it would be difficult to justify long-term bioassays with Compound C, since a limited number of workers are potentially exposed to an unmarketable product, namely an intermediate. In other words, the effort and cost for controls should be put in proper perspective.

The discussion following the presentation of this hypothetical test case was divided into two camps: those who favored doing long-term bioassays and those who felt additional short-term carcinogenic tests would suffice. Of course, there is no "right" answer to this problem.

II. DISCUSSION*

Compound C — Intermediate in a plant production step; modest exposure only to workers in the plant; related to a known carcinogen; negative Ames test; negative *Drosophila*; negative dominant lethal; positive mammalian cell transformation assay.

Clive — Is this a real situation?

Butterworth — They're closely related to real cases. I tried to set down postulated properties that reflect the actual kinds of situations that one does encounter, but in this instance the case is hypothetical. Were not talking about any specific compound.

Hite — I think they could be real. We have Compound C, it's an intermediate in a plant production step, with only moderate exposure to workers in the plant, related to a known carcinogen, and it's negative in the Ames test, *Drosophila*, and dominant lethal tests. However, it is positive in a mammalian cell transformation assay. We have a situation where this compound is negative in mutagenicity tests, but it is positive in a short-term carcinogenicity test. I am going to assume that each test is in itself valid and scientifically sound and can, in fact, be reproduced if need be. Again, we have a single carcinogenicity test that is positive.

We should first discuss the potential exposure of Compound C to the workers. This is important because of the positive result in the transformation assay. We should look into protecting the workers using good engineering controls and air handling systems. Let's say that it is the kind of material that could be inhaled by those people directly responsible for its manufacture. These people should be required to wear proper protective clothing, depending upon the nature of the chemical. I assume the chemical properties are known. The environment should be monitored and discussions should be arranged with the plant physician so that a medical surveillance program could be instituted. These are not very costly items. Since the compound is an intermediate, it probably will not be a public health problem. Therefore, the efforts spent on this kind of a compound should not warrant long-term bioassays. A long-term bioassay might tell us whether or not it is a carcinogen, but I'd like to hear your opinion as to whether or not we should go to bioassays. I personally don't think it is necessary at this time. There are a very limited number of people who have potential exposure.

Butterworth — I think you've just broken some new ground here. Youre saying then that even though we don't have the 2-year tests, on the basis of the positive mammalian cell transformation assay we should start physically protecting people in the plant. Do you agree with that?

Thilly — In terms of widely held beliefs about mutation and cancer, negative results in several mutagenicity tests and positive results in an in vitro transformation assay might be suspicious. However, there are examples of non-mutagenic carcinogens (asbestos) and noncarcinogenic mutagens (bromodeoxyuridine). I agree with Mark (Hite) that when we consider this situation from a practical standpoint it would seem that effective engineering controls, not more extensive testing, is the approach indicated.

Hite — I would like to propose to the group that we go back to one of the given statements, namely that it is related to a known carcinogen. Perhaps we ought to take the known carcinogen and put that into the battery of three mutagen tests that were done, namely the Ames test, *Drosophila,* and dominant lethal tests. In addition, I propose that not only the known carcinogen, but also Compound C be studied in some DNA tests, because the Ames test does have its limitations as we heard yesterday. *Drosophila* is the only whole-animal genetic test that was done with this particular compound and the dominant lethal test has questionable sensitivity.

Golberg — I think we should examine the evidence that it is a known carcinogen. Often mistakes were made, and this has been mentioned before; hence, one needs to evaluate the data very carefully.

Williams — I think one of the possibilities here is that the negative mutagenesis test results were due to the fact that whatever activation that might have taken place in the added microsomes produced metabolites so short-lived that they didn't reach the target organism. Thus,

---

* The transcript presented here is an edited version of the original. Some sections have been omitted and wording is not necessarily verbatim. Discussion leader is Mark Hite.

transformation in the mammalian system may have occurred as a result of cellular metabolism. But if you are going to go on to perform a DNA repair or damage assay, let's just say for the sake of discussion, I would predict that it would be negative in Swenberg's system but that it might be positive in the primary hepatocyte system in which metabolism takes place intercellularly. Therefore, I think you would want to be sure to include an assay with this potential in your evaluation.

Hite — Any rebuttal?

Swenberg — I'm going to surprise you. The way I tend to approach this would be to take that known carcinogen that Compound C is related to and test it in this same battery of tests. If it comes out positive in the transformation and negative in the others, I'm afraid I wouldn't run it in DNA damage and repair systems. I would proceed to look into the engineering aspects of Compound C production, with the aim of adequate containment. When you have this much evidence, you probably don't need a test involving 400 or 500 animals to show whether Compound C is or is not a carcinogen. Nevertheless, if at all possible, I'd go ahead and put the compound in some real animals and see if they get real tumors, and I wouldn't bother with the DNA damage and repair.

Schwetz — I'd like to follow up with some questions that the plant supervisors would certainly have: How much can you be exposed to? How tight do you have to button up at this point? Do I have to make them wear respirators 24 hr a day while they're working in this area? How often do I have to give them a physical examination? What is OSHA going to do about it? I think without whole animal long-term data of some nature, it becomes very difficult to know how much those people can be exposed to. While we sit in a laboratory saying that this compound is a suspect carcinogen, the workers are out there being exposed to it in quantities where we have no idea whether it represents a hazard or not.

Williams — Well, after all, one of the purposes of the short-term tests is, if possible, to save you from going into the long-term animal studies. I would still suggest that you should exploit the potential of the short-term tests fully before doing what Jim (Swenberg) suggested. Now, in the approach that I brought up, the alternatives would be as follows: if the compound were positive in the primary hepatocyte DNA repair test, then I think you would have to come to the conclusion that intracellular activation did occur and that the carcinogen is genotoxic. In that case, procedures to eliminate exposure should be instituted. If it's a negative, then you would have to consider that you may have a compound that is transforming by virtue of some epigenetic, hormonal, or promotional event. You would then require the animal studies if you wanted to establish the mechanism of action. But, I suppose if you got the cell transformation data at the beginning, there's nothing much that's going to get you off the hook.

Participant — I presume you've probably already taken some precautions, maybe not as tight as you would for a known carcinogen, but some effort to limit and avoid exposure. It's important that you review what you've already done every time you get another piece of bad news and see if there's a little more tightening up you can do. I suspect you should take another look to make sure that Compound C is not getting into the final product, perhaps try to increase the analytical sensitivity to have greater assurance that Compound C wasn't in there. Then, my feeling would be that if were concerned with those employees, we'd go right into the 2-year animal studies.

Hite — We have to have a hazard evaluation here. In other words, what is the likelihood of exposure of people to the compound. Institution of engineering controls along with proper medical surveillance is one reasonable approach.

Participant — When it comes to the employees or people who start asking questions, the only answer you can provide is that we've come up with what we think is a good assay, and explain it as straightforwardly as you can. They don't want speculation, they want an answer.

Participant — I dont really think we've answered Bern's (Schwetz) question about how much we have to tighten things up unless we know what the potency is in a whole animal system.

Participant — I think in the proposed Chemical Industry Institute of Toxicology document and in these examples we have inappropriately minimized written precautionary statements. I think early in production, in safety manuals, and in labeling, there should be early precautionary statements, not using the words "carcinogenic" or "mutagenic," but insuring that employees develop appropriate handling respect for these chemicals. I think it is the responsibility of toxicologists to insure that these written statements as guidelines get incorporated into practice very early so that proper handling training goes with the new chemical, so that we don't have to change work habits later. In benzene surveys we see that employees are scared. They're asking how come benzene is so toxic all of a sudden, what's the big hazard, what happens to it? Reeducating people who quite possibly don't have much education to begin with is a very difficult condition.

Longstaff — When you're talking about industrial hygiene, there is a significant difference between what is practicable and what is possible. It may be possible at tremendous expense to approach zero exposure. How do you decide what is reasonable at this stage?

Participant — This is why it's so urgent to get into whole animal studies as soon as possible.

Kilian — There's no such thing as absolute engineering control. I know a situation where an animal carcinogen was used in a process, and there was no worker exposure, but there was an accident where the carcinogen was blown over half a block area. I think you have to keep that in mind; that's why I think you've got to go all the way on the animal bioassay program.

Williams — There is something that's not clear to me concerning the handling of this compound. If you have taken some precautions to limit exposure because it's transforming cells, and then you do the animal study that reveals it to be carcinogenic, is there something more that you would now do that hadn't already been done? I mean, have you just employed some minimal procedures that you would now tighten up further, or would the compound be abandoned?

Legator — Isn't this the history of most industrial compounds; when they find more hazards we tighten up on regulations?

Participant — We hope that the animal data will give us some indication of the carcinogenic potency of the compound.

Participant — We must go step by step.

Williams — How would the workers respond to this now if you told them that it was known 3 years ago that this agent transformed cells in vitro, but until that was confirmed we allowed moderate exposure to the compound and now were going to tighten up.

Participant — That's not what they said at all. They said they were going to tighten it up as well as they could, but they would not make absolute commitments on a totally exposure-free environment until more data were available. They want to know whether the compound elicits tumors at 50 ppb or ppm or at 500 or 5000 ppm, or what. If you get some further bad news about your compound, I think that you should take a further look and see whether you can improve safety procedures. You may be able to make some substantial improvements. You may decide with positive animal data that no matter what it costs, were going for absolute containment.

Williams — Well, if you've agreed that you are handling it as a carcinogen as far as exposure to the workers is concerned, then what more safety measure is there left at the end of the bioassay?

Participant — It shouldn't be too much, I agree with you there.

Flamm — It sounds as if they're asking questions relating to potency, as might be the case in some of the vinyl chloride studies, where they're looking at a variety of different concentrations in the mouse. Unfortunately we're finding that quite low concentrations are producing cancer in the mouse, and this fact has had impact on the action levels that OSHA has been promulgating. I think they've gotten it down to something like 1 ppm at the present time, but they're still able to demonstrate carcinogenicity in the mouse at this level. The goal is zero; obviously, one cannot achieve that goal. Even the Environmental Defense Fund, according to its General Counsel, feels that while the animal studies would suggest that there is still considerable hazard at 1 ppm, at least it's the devil we know, as opposed to all those other polymeric monomeric situations that we don't know. So what they're going to try to do is to keep tightening up on controls as is feasible rather than to ban these processes in polyvinyl chloride manufacture, which would cause reliance on other products that may be as bad or worse, but we just don't know about because the studies haven't been done. It seems to have impact primarily on the perception of degree of hazard from these different exposures so that it would seem to me appropriate to perform pharmacokinetic and metabolic studies that would build some bridges from the animal model to man and so we can generate more confidence with regard to the significance of the potency findings in the mouse in comparison with potential potency in humans. It may mean, too, that we need to know more about the kinds of adducts that form on the DNA with metabolites of vinyl chloride so we can do dosimetric type calculations based on knowledge of of the pharmacokinetics and the amount of material that binds DNA in the different species and so forth and so on. I don't think the question that is being asked here is so much is it or is it not a carcinogen, but what does it look like in a reasonably respectable animal model system and what information can one derive in terms of hazard from the point of view of the population at risk?

Williams — What you're suggesting in doing a dose-response curve is a much larger undertaking.

Flamm — The question that's being asked has to be rather well defined in that one has to look at whether the implementation of the experimental work is going to be adequate to answer the question.

Williams — Now you're talking about going to dose-response, which is a very big undertaking and you're going to want to come out at the end of it saying that the precautions we took over these 3 years were alright in terms of the dose-response data that we now have. I really come back to the point at the beginning that if further in vitro studies, such as the DNA repair assay, were performed and found to be positive, then you might as well consider it a carcinogen and implement the full precautions right at that point.

Flamm — There is no such thing as "full precautions" right at that point; it's generally an evolving kind of situation. I don't know whether it is in this case; this is a generic example. It's like getting the aflatoxin levels in peanut butter down lower. You work as hard as you can and one manufacturer does a better job than another manufacturer and you try to set the level between the two and that encourages it to go down further and further. It's one of the reasons there are laws against adulterations to achieve action levels and residues in food, because if it weren't for these differences between manufacturers and processors, regulators would have no way of knowing what's possible. They keep trying to achieve what is possible in the interests of occupational or public health.

Henderson — You raised the question of what the unions may do. When you have a plant in a high unemployment area and it's about to get shut down because the limits of exposure get set so low, the union may object to those limits being put that low. I saw that happen in vinyl chloride manufacture. The union wrote a letter objecting to having limits set too low. So the employees who belong to unions are also realistic enough to recognize, as you have said, that they'd rather live with hazardous conditions they understand than be unemployed.

Williams — I for one wouldn't let them make that decision.

Henderson — We don't know whether the 1 ppm allowable on vinyl chloride is going to absolutely protect them and yet we're still operating in that business.

Participant — I would simply recommend a well-designed, long-term study to provide us with valuable information necessary in evaluating the actual hazard.

Soares — There's one other point I'd like to bring up, and this applies to looking at the potency of the carcinogen, and for that matter, any kind of 2-year animal study. The strain of animal that you're going to use in a particular study is extremely important because, in fact, it has been shown that the genotype of the animal is going to control to a large extent its response to the compound you're exposing the animal to. For instance, in the mouse ethylene oxide is extremely toxic to one strain of mouse and yet another unrelated strain of mouse can handle ethylene oxide without any problem. The same applies to carcinogenicity studies. The question that might come up is: if the compound is not carcinogenic in one strain, then what about in another strain?

Index

INDEX

A

N-Acetoxy-AAF, see N-Acetoxyacetylaminofluorene
N-Acetoxyacetylaminofluorene, 15, 18, 82
Acetylaminofluorene, 15, 30, 32, 82, 83
AE, see Alkaline elution
Aflatoxin B_1, 17, 30, 32, 83
   mutagenicity, 31
8A6, see 8-Azaguanine
Alkaline elution (AE), 79, 83—84
Alkaline sucrose gradient (ASG), 79, 82
Alkanesulfonate derivatives, mutagenicity and cytotoxicity, 48
Alkylating agent, mutagenicity, 49
7-Alkylguanine, 78
Alkylsulfate derivatives, mutagenicity and cytotoxicity, 48
Ames *Salmonella*/microsome assay, 4—8, 11—12, 22—26, 97, 117—119
9-Aminoacridine, 18, 30, 32, 35
2-Aminobiphenyl, 81
Ammonia, 82
Apurinic site, 78
Apyrimidinic site, 78
Arbinofuranocytosine, 15
ASG, see Alkaline sucrose gradient
Asparagine, 14
8-Azaguanine (8A6), 14, 19, 40
Azaserine, 84
Azoxymethane, 83

B

BA, see Benzanthracene
Bacterial mutagenesis assay, see also specific assay systems, 3—8
   compound variability, 10
results, 10—12
variability, 8—10
   complete quality control, 9—10
   strain quality control, 8—9
Base-pair substitution mutation, 33—34
Battery system, 90
Benzanthracene (BA), 16, 17, 81
Benzanthracene-*cis*-dihydrodiol, 16
Benzanthracene-*trans*-dihydrodiol, 16
Benzanthracene-k^c-epoxide, 16
Benzanthracene-k-phenol, 16
Benzidine HCl, 83
Benzo(a)pyrene, 17, 30, 32, 81
5,6-Benzoquinoline, toxicity, 31
Bromodeoxyuridine (BUdR), 15, 16, 18
7-Bromodimethlbenzanthracene, 16
7-Bromo-12-methlbenzanthracene, 16
BUdR, see Bromodeoxyuridine

C

Caffeine, 15
Carcinogenic process, 78
Chain termination mutation, 34
Chinese hamster lung cell (V79), 14
   in vitro mutagenesis, 15—17
Chinese hamster ovary cell (CHO), 14
   characteristics of, 41
   in vitro mutagenesis, 15
Chlordane, 81
Chlorodeoxyuridine, 18
Chloroform, 84
CHO, see Chinese hamster ovary cell
CHO/HGPRT assay, 40
   colony-forming ability, 51
   conclusions, 50—53
   correlation of mutagenicity, 52
   effect of cell number, 43
   expression time of mutant phenotype, 42
   genetic basis, 45
   materials and methods,
      host mediated, 43
      microsome activation system coupling, 41—42
      selective system, 41
   mutation induction, 51
   results
      dose response relationship, 44—47
      dosimetry, 47
      genetic origin of mutation induction, 43—44
      mutagenecity of physical and chemical agents, 48—50
      structure activity relationships, 48
   validation, 50
Chrysene, 16
Cultured mammalian cell transformation system, see also CHO/HGPRT assay, 55—56
   conclusion, 63—64
   epithelial cell, 63
   hamster embryo cell, see also Chinese hamster ovary cell, 56—57
      BHK$_{21}$ cell line, 58—59
      transplacental assay, 57
   human cell, 63
   mouse cell lines, 62—63
      BALB/3T3, 59
      C3H/10T1/2, 59—60
      M2, 60—61
   rat embryo cell, 61
   virus and chemical systems
      hamster embryo cells, 61
      mouse embryo cell lines, 62—63
      rat embryo cell lines, 61—62
Cyclohexamide resistance, 14
Cyclophasphamide, 81, 84

D

DBA, see Dibenzanthracene
Dibenzanthracene (DBA), 16
1,2,3,4-Dibenzanthracene, 30, 32
Dibenz[a,c]anthracene, 17
Dibenz[a,h]anthracene, 17
Dibenzanthracene-*cis*-dihyrodiol, 16
Dibenzanthracene-*trans*-dihydroldiol, 16
Dibenzanthracene-k-epoxide, 16
Dibenzanthracene-k-phenol, 16
Dieldrin, 81
Diethylnitrosamine, 17, 83
Diethylsulfate, 29
3-2'-Dimethyl-4-aminobiphenyl, 82
7,12-Dimethylbenzanthracene (DMBA), 16, 17, 30, 32, 83
7,12-Dimethylbenzanthracene-k-epoxide, 16
1,2-Dimethylhydrazine, 82, 83
Dimethylnitrosamine, 15, 17, 29, 32, 82, 83
Dimethylsulfate, 29
Dimethylsulfonate, 32
Dipentylnitrosamine, 17
Diphenylnitrosamine, 17
Dipropylnitrosamine, 17
DLA, see Dominant lethal assay
DMBA, see 7,12-Dimethylbenzanthracene
DNA, pathways for damage and repair, 79
DNA damage and repair assay, 77—79, 120—121
    divergent results of, 81
    prediction of carcinogenic potential, 80
    validation study
        in vitro, 79—81
        in vivo, 81—85
DNA modification assay, 8
    host mediated scheme, 9
Dominant lethal assay (DLA), 72
Drosophila, in vivo mutagenicity testing with, 68, 110

E

Escherichia coli
    lac operon, 32
    streptomycin independence, 22
Ethyl carbamate, 84
Ethylmethanesulfonate, 15, 16, 17, 18, 29, 32, 83
    dosimetry of induced mutagenesis, 47
    effect on colony-forming ability and mutation induction, 46
    mutation induction mediated by, 44—47
Ethylnitrosourea, 82, 83

F

Federal Insecticide, Fungicide, and Rodenticide Act (FIFRA), 110
FIFRA, see Federal Insecticide, Fungicide, and Rodenticide Act
Fluorodeoxyuridine, 15
Forward mutation assay, 21, 29—30
    biochemical basis, 27

Frameshift mutation, 34
Fructose, 14

G

Gamma radiation, 16
Germ cell assay
    dominant lethal assay (DLA), 72
    heritable translocation assay (HTA), 73
    specific locus test (SLT), 74—75
    sperm morphology, 73—74
Gene locus mutation assay, 13
    bacterial, 19—26
    conclusion, 36
    drug metabolism, 26
    mamillian cell, 13—19
        phenotypic lag, 19—21
    structure versus mutagenicity, 26, 32—35
Genetic code, 33

H

Haemophilus influenzae
    novobiocin resistance, 21
Hamster embryo cell, see also Chinese hamster embryo cell, 56—57, 61—62
    BHK_{21} line, 58—59
    transplacental assay, 57
Heat, 18
Heritable translocation assay (HTA), 73
HTA, see Heritable translocation assay
Heterocyclic nitrogen mustard, see ICR
Human fibroblast
    fetal lung, in vitro mutagenesis, 18
    secondary culture, in vitro mutagenesis, 18
    strain HFIO, in vitro mutagenesis, 18
Human lymphoblast
    line PGLC-33H, in vitro mutagenesis, 18
Hycanthonemethanesulfonate, 17
N-Hydroxyacetylaminofluorene, 15, 83
Hydroxylamine, 15
1'-Hydroxysafrole, 81
Hydroxyurea, 15
Hypoxanthine-guanine phosphoribolsyltransferase (HGPRT), 17, 40

I

IA3, 17
IA4, 17
IA5, 17
IA6, 17
ICR compounds, see also specific types
    mutagenicity and cytotoxicity, 48, 49
ICR-191, 15, 17, 18, 30, 32
ICR-372, 17
In vivo mammalian test systems, see also specific systems
    germ cell assay, see Germ cell assay
    mutagenicity testing, 68—69

short term, 67—68
somatic cell assay, see Somatic cell assay
Iododeoxyuridine, 14, 18

L

Lac operon, in *E. coli*, 32
Large deletion mutation, 35
Lead acetate, 81
Lucanthone HCl, 17

M

MAM, 82
Mammalian cell mutation system, see CHO/HGPRT assay
Mammalian cell transformation system, see Cultured mammalian cell transformation system
Metallic compounds, mutagenicity, 49
7-Methylbenzanthrene, 16, 17
Methyl-t-butylnitrosamine, 17
MCA, see 3-Methylcholanthrene
3-Methylcholanthrene (MCA), 16, 17
3-Methylcholanthrene-*cis*-dihydrodiol, 16
3-Methylcholanthrene-*trans*-dihydrodiol, 16
3-Methylcholanthrene-k-epoxide, 16
3-Methylcholanthrene-k-phenol, 16
Methylenedimethanesulfonate, 18
Methylmethanesulfonate, 15, 18, 29, 32, 83
Methynitrosoguanidine, 15, 17, 18, 29
N-Methyl-*N*-nitro-nitrosoguanine, 32, 82, 83
Methylnitrosourea, 15, 18, 82, 83
Methylnitrosourethane, 15, 18, 29, 32, 82
Methylphenylnitrosamine, 17
5-Methyltryptophan, 21
Microbial assay, see Bacterial mutagenesis assay
Micronucleus test, 69
Mitomycin C, 82
Mouse cell line
    BALB/3T3, 59
    C3H/10T1/2, 59—60
    M2, 60-61
Mouse lymphoma cell
    L5178T, 14
        in vitro mutagenesis, 17
    P388, 14
        in vitro mutagenesis, 18
    S49, 14
        in vitro mutagenesis, 18

N

Nitrogen mustard, 18
4-Nitroguinoline-1-oxide, 82, 83
Nitrosamines, mutagenicity, 49
Nitroso compounds mutagenicity and cytotoxicity, 48, 49
Nitrosoguanidines cytotoxicity and mutagenicity, 49
N-Nitrosohexamethyleneimine, 83
Nitrosomethylpiperazine, 17
Nitrosomorpholine, 17
Nitrosopyrrolidine, 17
Nitrosoureas, cytotoxicity and mutagenicity, 49
Novobiocin resistance, 21

O

Ouabain resistance, 14

P

Phenanthrene, 17
Phenotypic lag, 19—21
Polycyclic hydrocarbons, mutagenicity, 49
Proflavin, 30, 32
β-Propiolactone, 29, 32, 83
Pyrene, 16

R

Rat embryo cell, 61
Recessive spot test, see Somatic cell recessive spot test
Reverse mutation assay, 21, 27—29
    agar incorporation method, 5
    gradient method, 6
    modified fluctuation method, 7
    spot and fabric methods, 4

S

Salmonella typhimurium, see also Ames *Salmonella*/microsome assay, 97, 98
    background mutant fraction, 28
    forward and reverse assays, 32
    test, see Ames *Salmonella*/microsome assay
SCE, see Sister chromatid exchange
Short-term testing, see also specific assays
    battery vs. Tier system, 90—91
course of action
    negative response, 101
    positve response, 100—101
    degree of development, 95
        correlation of results, 97
        metabolic action in vitro, 96
    evaluation, 99—100
    false negative responses, 91
    genetic considerations, 67—68
    hypothetical case A, 105—116
    hypothetical case B, 117—121
    hypothetical case C, 123—127
    resources and expertise, 91—95
    validity, 97, 120
        mutagenicity of materials, 99
        NCI studies, 98
        standardization, 98—99
    value, 89—90

Sister chromatid exchange (SCE), 70—71
SLT, see Specific locus test
SMA, see Sperm morphology assay
Somatic cell assay
   cell cytogenesis, 69—70
   micronucleus test, 69
   sister chromatid exchange (SCE), 70—71
   recessive spot test, 71—72
Somatic cell cytogenesis, 69—70
Somatic cell recessive spot test, 71—72
Specific locus test (SLT), 74—75
Sperm morphology assay (SMA), 73—74
Streptomycin independence, 21
Streptozocin, 84
Sunlight, lethality and mutagenicity, 52
Syrian hamster embryo colony assay, 57

T

Thymidine, 14
Thymidine kinase (TK), 14
Tier system, 90
6TG, see 6-Thioguanine
6-Thioguanine (6TG), 14, 19, 40
   mutation to resistance, response curve, 23, 24

TK, see Thymidine kinase
Toxic Substances Control Act (TSCA), 110
Triflourothymidine, 15
TSCA, see Toxic Substances Control Act

U

Ultraviolet light, 15, 16, 18

V

V79, see Chinese hamster lung cell
Vinyl chloride, 126—127

X

Xanthine-guanine phosphoribosyltransferase (XGPRT), 25
XGPRT, see Xanthine-guanine phosphoribosyltransferase, 25
X-ray, 15, 16, 17, 18